Energy Industries and Sustainability

宝库山精选：能源产业与可持续性

Digital editions

Energy Industries and Sustainability is available through most major ebook and database services (please check with them for pricing). Special print/digital bundle pricing is also available in cooperation with Credo Reference; contact Berkshire Publishing (info@berkshirepublishing.com) for details.

For information, contact:
Berkshire Publishing Group LLC
122 Castle Street
Great Barrington, Massachusetts 01230-1506 USA
www.berkshirepublishing.com
Printed in the United States of America

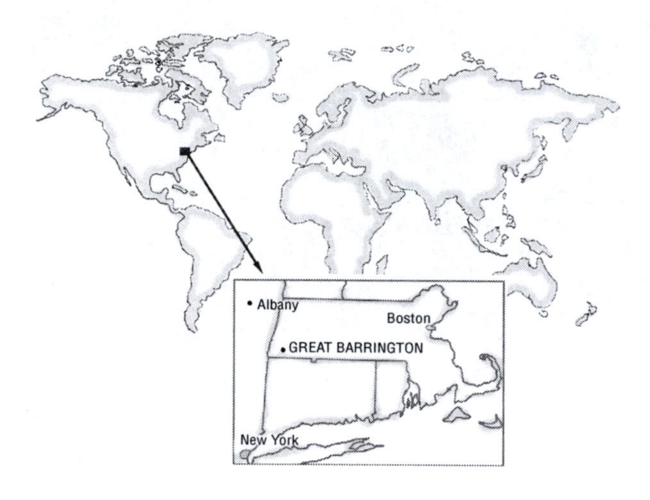

Library of Congress Cataloging-in-Publication Data

Energy industries and sustainability / general editor, Ray C. Anderson ; editors, Daniel S. Fogel et al.
pages cm.—(Berkshire essentials)
Includes bibliographical references and index.
ISBN 978-1-61472-990-7 (pbk. : alk. paper)—ISBN 978-1-61472-968-6 (ebook)
1. Energy industries—Environmental aspects. 2. Sustainability. I. Anderson, Ray C., editor of compilation. II. Fogel, Daniel S., editor of compilation.
HD9502.A2E54399 2013
333—dc23 2013032126

BERKSHIRE 宝库山

Essentials

Energy Industries and Sustainability

宝库山精选：能源产业与可持续性

General Editor: Ray C. Anderson. *Editors:* Klaus Bosselmann, Robin Kundis Craig, Daniel S. Fogel, Sarah E. Fredericks, Lisa M. Butler Harrington, Willis Jenkins, Chris Laszlo, John Copeland Nagle, Bruce Pardy, J.B. Ruhl, Oswald J. Schmitz, Shen Lei, William K. Smith, Ian Spellerberg, Shirley Thompson, Daniel E. Vasey, Gernot Wagner, & Peter Whitehouse

About *Energy Industries and Sustainability*

Energy Industries and Sustainability, a Berkshire Essential, covers the exploitation of energy resources—such as coal, petroleum, and wood—and the innovations that can provide the energy we need for a cleaner, safer, and more sustainable future. Forty expert authors explain concepts such as "materials substitution" and the "polluter pays principle" and examine the industries and practices that bring us energy from the sun, water, and wind. This concise handbook offers a broad view of positive steps being taken to make responsible energy use a priority around the globe, and is designed for use in classrooms at the high school and college level. The book will be helpful to engaged citizens as well as to business people, policy makers, and environmental professionals. Controversial topics such as nuclear power and fracking are explained clearly and impartially, with a view to promoting thoughtful discussion and informed decision-making.

THE **BERKSHIRE** *Essentials* SERIES

Berkshire Sustainability Essentials, distilled from the *Berkshire Encyclopedia of Sustainability*, take a global approach to environmental law, energy, business strategies and management, industrial ecology, and religion, among other topics.

- Religion and Sustainability
- Business Strategies and Management for Sustainability
- Ecosystem Services and Sustainability
- Energy Industries and Sustainability
- Energy Resources and Sustainability
- Environmental Law and Sustainability
- Finance and Investment for Sustainability
- Industrial Ecology

Distilled for the classroom from Berkshire's award-winning encyclopedias

BERKSHIRE ESSENTIALS **from the *Berkshire Encyclopedia of China* and the *Berkshire Encyclopedia of World History, 2nd Edition* also available.**

Contents

About Berkshire Essentials

For more than a decade, Berkshire Publishing has collaborated with a worldwide network of scholars and editors to produce award-winning academic resources on popular subjects for a discerning audience. The "Berkshire Essentials" series are collections of concentrated content, inspired by requests from teachers, curriculum planners, and professors who praise the encyclopedic approach of Berkshire's reference works, but who still crave single volumes for course use.

Each Essentials series draws from Berkshire publications on a big topic—world history, Chinese studies, and environmental sustainability, for instance—to provide thematic volumes that can be purchased alone, in any combination, or as a set. Teachers will find the insightful articles indispensable for stimulating classroom discussion or independent study. Students, professionals, and general readers all will find the articles invaluable when exploring a line of research or an abiding interest.

These affordable books are available in paperback as well as ebook formats for convenient reading on mobile devices.

Editors, Editorial Advisory Board, and Production Staff

Editors

The following people served as Editors for the sources of these articles (all from the *Berkshire Encyclopedia of Sustainability*):

From Volume 1, *The Spirit of Sustainability:* Willis Jenkins, General Editor, *Yale University;* Whitney Bauman, *Florida International University*

From Volume 2, *The Business of Sustainability:* Chris Laszlo, General Editor, *Case Western Reserve University;* Karen Christensen, *Berkshire Publishing Group;* Daniel S. Fogel, *Wake Forest University;* Gernot Wagner, *Environmental Defense Fund;* Peter Whitehouse, *Case Western Reserve University*

From Volume 3, *The Law and Politics of Sustainability:* Klaus Bosselmann, *University of Auckland,* Daniel S. Fogel, *Wake Forest University,* J.B. Ruhl, *Vanderbilt University Law School*

From Volume 4, *Natural Resources and Sustainability:* Daniel E. Vasey, General Editor, *Divine Word College,* Sarah E. Fredericks, *University of North Texas,* Lei Shen, *Chinese Academy of Sciences,* Shirley Thompson, *University of Manitoba*

From Volume 5, *Ecosystem Management and Sustainability:* Robin Kundis Craig, *University of Utah,* John Copeland Nagle, *University of Notre Dame,* Bruce Pardy, *Queen's University,* Oswald J. Schmitz, *Yale University,* William K. Smith, *Wake Forest University*

From Volume 6, *Measurements, Indicators, and Research Methods for Sustainability:* Ian Spellerberg, General Editor, *Lincoln University,* Daniel S. Fogel, *Wake Forest University,* Sarah E. Fredericks, *University of North Texas,* Lisa M. Butler Harrington, *Kansas State University*

Editorial Advisory Board

Production Staff

Project Coordinator
Bill Siever

Copy Editors
Linda Aspen-Baxter
Mary Bagg
Kathy Brock
Carolyn Haley
Barbara Resch
Elma Sanders
Chris Yurko

Editorial Assistants
Echo Bergquist
David Gagne
Ellie Johnston

Designer
Anna Myers

Introduction to Renewable Energy

The term renewable energy *refers to any energy source that is naturally occurring and abundant—for example, solar, wind, hydro, biomass, tidal, and geothermal. Although appealing because many forms of renewable energy are widely available and they do not rely on fossil fuels (and therefore are considered "cleaner"), they are still more expensive for producers and consumers. Putting a price on carbon emissions, however, may level the field between renewables and fossil fuels.*

The case for rapid growth in renewable energy is clear. Catastrophic climate change, as well as many other environmental, economic, and social factors, necessitates a transformation of the $5 trillion per year fossil fuel–based energy sector (Stuebi 2009).

Part of that transformation will involve cleaner ways of using fossil fuels. Another part entails making everything that uses energy more efficient. A third part comes under the broad heading of behavioral change and aims to decrease demand or limit demand increases. Fourth are far-reaching approaches to changing entire systems like transport, the way we commute, and travel. The fifth, and likely most significant, is renewable energy, producing much-needed energy in cleaner ways.

This chapter focuses on the policy question of how to bring about change in the electricity sector. The answer is twofold: price plays one important role, and the other is an approach to regulation that seeks to encourage wholesale change in how electricity is produced, delivered, and used.

Fast Growth

Many renewable energy technologies have been growing at double-digit rates for years and are expected to continue to grow at similar (or even faster) rates well into the future.

Individual chapters in this section describe the development of each renewable sector and its future prospects. Figure 1 on page 2 summarizes growth trends for wind, solar, and biomass through 2020, all of which are projected to continue to grow rapidly at least through then; figure 2, on page 3, summarizes the European Union's experience with renewable energy production.

Rapidly Declining Costs

An important element in the growth of renewable energy technologies is their cost relative to traditional fuels. It is often more expensive to install wind, solar, biomass, or hydro plants than a coal or gas plant producing the same amount of electricity. Once installed, however, renewable plants provide virtually free power, but average electricity prices from coal are still cheaper than most of the cleaner alternatives. Nonetheless, costs for the latter have been coming down quickly.

The holy grail is "grid parity"—the point at which renewables find themselves on equal footing with coal and gas. Some forms of onshore wind power have achieved this already. Others are fast approaching this point.

Shape of Things to Come

History demonstrates that the introduction of new technologies is often explosive rather than linear. Growth in technologies usually follows an S-shaped curve: slow start, rapid acceleration, and then tailing off at the end as the former innovations become ubiquitous. When successful, the market usually spreads these new technologies more widely and cheaply than had been predicted.

Renewable energy in general will clearly follow this pattern. One sign is the increase of patents for clean

Figure 1. Growth Potential of Renewable Energy Technologies

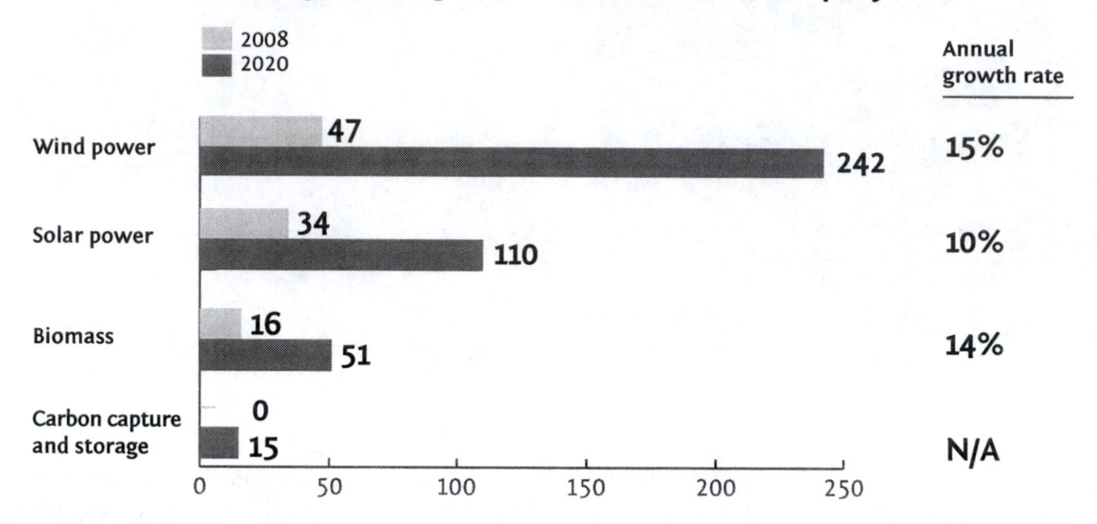

Source: McKinsey & Company (2009a, 41).

Note: Numbers represent billions per year in C (€).

The market for low emission energy technologies is expected to grow by 13 percent per year through 2020; wind power expenditures are expected to see the greatest increase, growing to €165 billion (approximately US$225 million) by 2020.

technology in Europe at a rate of roughly 10 percent annually between 1997 and 2007. The United States and Australia, which have not implemented strong climate policies as of early 2010, have experienced no such increase. To be sure, low-carbon innovations have begun in the United States, spurred by some state-level emission caps and the prospect of national-level controls. But only when the long-term economic signals are in place and innovation clearly pays will entrepreneurs and investors move into high gear.

Regulating Change

A key aspect of bringing about change in the power sector is by ensuring that the price of electricity includes the social costs of carbon pollution. Without proper regulation, emitting global-warming pollution is free—or even supported through fossil fuel subsidies. Putting a price on carbon would make renewable technologies more attractive vis-à-vis traditional fuels and hasten the transition to a low-carbon economy.

A market for carbon is the obvious way to bring this about, but creating one requires a limit on carbon emissions—a cap. And while it would be prohibitively complex and expensive to impose individual carbon limits for every business, it is relatively simple and cheap to turn the obligation to reduce carbon emissions into an opportunity. The fundamental task is to reward emissions reductions in the most flexible way.

The method for doing this is called, somewhat prosaically, "cap-and-trade." It should really be called "rewards for innovation." Cap-and-trade is an enormous, publicly structured Robin Hood program to take money from the inefficient and unimaginative and pay the efficient and innovative, without deciding beforehand who is in which group. It is the definition of "doing good by doing well" and will be a fundamental driver of change in the energy sector.

Rate Decoupling and Full Unbundling

Another important regulatory mechanism is the decoupling of electricity sales from a utility's profits. Traditional utilities, like all other industries, benefit when they sell more product. Their product is electricity. Hence they have little economic incentive to distribute it more efficiently or to motivate customers to conserve.

One way to cause utilities to change is to set renewable electricity standards or energy efficiency mandates. Sometimes these measures are justified, but as with any economic standard, this is an inefficient way to go about doing business. Utilities will act to meet the regulation but no more. Standards cap innovation and adaptation. They do not cap carbon emissions or any other form of pollution.

Figure 2. Electricity Generation Technology Learning Curves for the European Union, 1980–1995

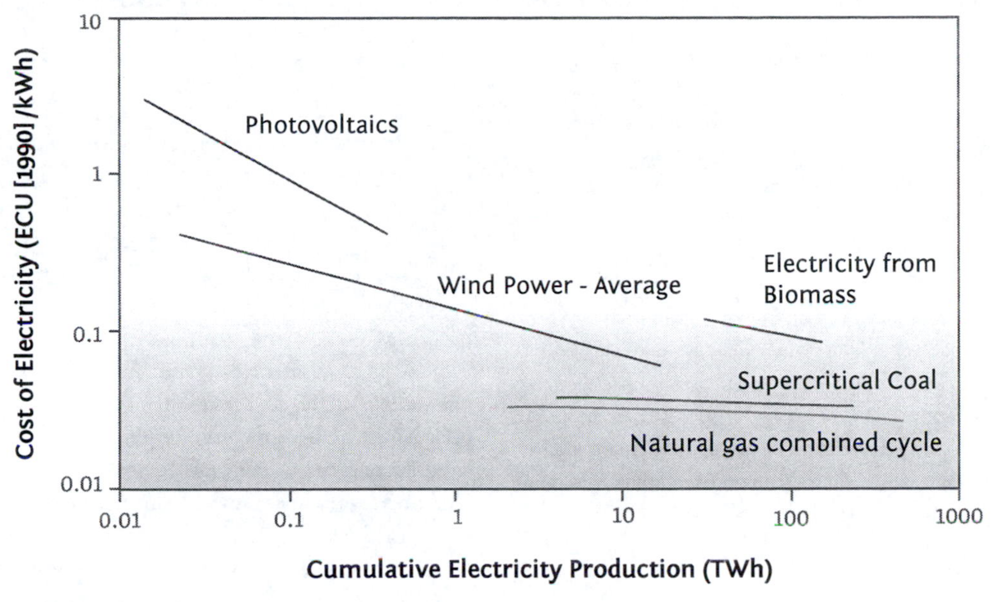

Source: Tam & Gielen (2007, 6).

From 1980 to 1995 the cost of generating electricity from renewable sources fell, while production from these sources increased. By 1995 the cost of wind power was less than 0.1 ECU (European Currency Unit) per kWh, and the European Union had produced more than 10 terawatt-hours of wind power (the ECU was replaced by the euro in 1999, at a value of 1 to 1).

Instead, utility profits ought to be decoupled from their sales. Breaking the profit-revenue link ensures that it is in the interest of the utility to motivate customers to use less of its product. As California showed when it adopted decoupling in the 1980s, breaking this link also enables another kind of decoupling that is ultimately most important—that between economic growth and emissions (Sudarshan and Sweeney 2008).

Decoupling is often only an intermediary step to an even better outcome: fully unbundled rates. Everyone gets paid for the exact services they are providing, and all services are clearly listed on the bill. Transmission is no longer part of the usage charge but is listed separately, and additional services and maintenance are assessed when needed instead of being lumped into the electricity rate.

That changes the equation for grid parity quite significantly. Individual solar panels suddenly become much more cost-effective, as they avoid the costs of using transmission through the grid. Full unbundling puts the costs of traditional energy in plain sight and makes distributed generation all the more attractive.

Distributed generation makes households not only consumers but also suppliers of energy—selling electricity back to the grid if they produce more than they use. Decoupling and full unbundling will help facilitate the large-scale deployment of renewable energy technologies like solar panels on rooftops or small geothermal facilities. Power in this kind of distributed system will frequently be used close to where it is produced. This will save much of the 10 percent of electricity that is currently lost during transmission, often across hundreds of miles of high-voltage cables, from plant to end user (USDOE 2009).

Smart Grids

A price on carbon, even with rate decoupling or full unbundling, is not enough to cause sustainable, systemic change in how power is produced. The complexities in the way we produce, distribute, store, and deliver energy demand a more holistic approach.

In many ways, the most exciting changes in the power sector will not come from technological breakthroughs in solar, wind, biomass, hydro, and others—as important as they may be—but through transformation of the entire system. The grid itself, the so-called smart grid of the near

What's the Difference Between a Gigawatt and a Gigawatt-Hour?

These terms often appear together, and they may seem similar. But understanding the distinction can help illuminate discussions about generation and consumption of electricity. Hearing that a new nuclear power plant has a 4 gigawatt capacity, for instance, is meaningless without something to compare the figure to.

The first thing to know is that *energy* is defined as the ability to do work, while *power* is energy consumed per unit of time. No matter if you drive a gas guzzler or a moped, the gasoline used has the same energy per liter or gallon of gasoline. A sports car, however, uses much more power than a more efficient vehicle, which allows it to go from point A to point B at a much faster rate.

What is a gigawatt?

A watt is a measure of power, the rate at which energy is converted to electricity. A 100 watt incandescent lightbulb uses 100 watts to light a room, while a compact fluorescent light or lamp (CFL) uses only 25 watts to provide a similar amount of light. One billion watts equals 1 gigawatt (GW); in 2006 this was enough to power 780,000 average US homes at average power in a given instant. (This calculation is based on data reported by the United States Energy Information Administration [EIA].)

How much power does it represent?

A gigawatt is enough to power in a given instant:

- 10 million 100-watt lightbulbs
- 561,000 average households in the relatively warm state of Tennessee
- 1,257,000 average Vermont households (mostly due to the relative lack of air conditioning used in the cold northeastern state)
- 1,967,000 average households in the United Kingdom
- 3,290,000 average Beijing households (or 7.2 million average low-income homes in Beijing)
- over 50,000,000 homes in areas of rural western China

Why the variation?

The local climate, the physical size of a household, the appliances (number, type, and efficiency), and the extent to which electricity is used for heating, cooling, heating water, and cooking (i.e., consumer behavior) can all influence the demand of a household. Estimates are calculated from average monthly or annual electricity consumption for each household type. Power consumption in a three-bedroom US household might vary from 0.5 kilowatts (at night) to 4 kilowatts or more (during peak hours); taken in aggregate, this can lead to large fluctuations in demand. In the province of Ontario, Canada, for example, demand on 4 March 2010 varied from about 1.4 GW at 3 a.m. to over 1.9 GW at 8 p.m. The province's record peak of 27 GW occurred during the summer of 2006—approximately fifteen times the average of the aforementioned day in March 2010 (IESO 2010).

Because power companies need to build enough capacity to cover peak demand (capacity that goes unused a majority of the time) or purchase power at a premium from other companies, these fluctuations mean greater cost to the consumer. Smart meters (meters that track consumption over time) allow producers to charge higher prices for peak consumption, as well as allowing consumers to keep track of how much energy they're using. Conventional meters—which are still widely used—only track total consumption, which means that producers must average peak power costs into the base price and consumers have no idea how much energy they're consuming at any given time. By reducing electricity usage during peak hours (for instance by doing laundry at night), consumers can help lower the costs of generating power.

What is a gigawatt-hour?

A gigawatt-hour (GWh), meanwhile, is a measure of *energy* equal to one gigawatt generated for one hour. Gigawatt-hours (as well as megawatt-hours and kilowatt-hours) are used because, generally speaking, it is not practical to measure how much

(Continued)

energy is used in a given second. As stated above, a gigawatt-hour is enough to power 561,000 Tennessee households for one hour.

The following are equivalent to 1 GWh; in the case of fuels (such as gasoline and liquid propane in the list below), 100 percent conversion to energy is assumed. (When a fuel is burned, the heat energy produced by combustion must be converted into electricity using a turbine, steam, or other means. Heat transfer and friction make perfect conversion impossible; a conversion rate of 38 percent of available heat energy to electricity is typical of modern fuel-powered generation systems [USEIA 2010a, 54].)

One gigawatt-hour is equivalent to:

- 1,000 megawatt-hours (MWh)
- 1 million kilowatt-hours (kWh)
- 3.41 billion British thermal units (Btu)
- 3.6 trillion joules (J)
- 860.4 billion calories (Cal)
- 86 tonnes of oil equivalent (toe)
- 123 tonnes of coal equivalent (tce)
- 141 cubic meters of liquid propane
- 97,124 cubic meters natural gas
- 103,866 liters gasoline
- 30,000 cubic meters hydrogen

Therefore if you consume an extraordinarily large hamburger (with condiments) containing 860 billion calories (about 1.5 billion times the size of a typical hamburger), you would then need to burn off a gigawatt-hour's worth of energy—a lot of time at the gym. Assuming you could jump rope at seventy or so skips per minute, this would take approximately 170,000 years.

David GAGNE and Bill SIEVER

Sources:

Department for Business Enterprise and Regulatory Reform. (2007). *Energy trends December 2007* (p. 24). Retrieved January, 15, 2010, from http://www.berr.gov.uk/files/file43304.pdf

Fit Watch (2010). Fast foods, hamburger, regular, double patty, with condiments. Retrieved March 11, 2010, from http://www.fitwatch.com/phpscripts/viewfood.php?ndb_no=21111&descr=Fast%20foods,%20hamburger,%20regular,%20double%20patty,%20with%20condiments

Gao, Peng, & Luo, Guoliang. (2009). Problems in development of electrical power in rural china. Retrieved March 5, 2010, from http://www.ccsenet.org/journal/index.php/ass/article/viewFile/4545/3878

Gong, John. (2010, January 5). Hike power prices for business, not families. Retrieved January 15, 2010, from http://www.shanghaidaily.com/article/print.asp?id=424713

International Energy Agency. (2010). Unit converter. Retrieved January 15, 2010, from http://www.iea.org/stats/unit.asp

Independent Electicity System Operator (IESO). (2010). Ontario demand and market prices. Retrieved March 5, 2010, from http://www.ieso.ca/imoweb/siteShared/demand_price.asp

National Energy Board, Canada. (2010). Energy conversion tables. Retrieved January 15, 2010, from http://www.neb.gc.ca/clf-nsi/rnrgynfmtn/sttstc/nrgycnvrsntbl/nrgycnvrsntbl-eng.html#s4ss3

United States Department of Energy (USDOE). How compact fluorescents compare with incandescents. Retrieved March 11, 2010, from http://www.energysavers.gov/your_home/lighting_daylighting/index.cfm/mytopic=12060

United States Energy Information Administration (USEIA). (2009). Table 5: US average monthly bill by sector, census division, and state 2006. Retrieved January 15, 2010, from http://www.eia.doe.gov/cneaf/electricity/esr/table5.html

United States Energy Information Administration (USEIA). (2010a). *Electric power annual 2008*. Retrieved March 3, 2010, from http://www.eia.doe.gov/cneaf/electricity/epa/epa.pdf

United States Energy Information Administration (USEIA). (2010b). Table 5: US average monthly bill by sector, census division, and state 2008. Retrieved February 12, 2010, from http://www.eia.doe.gov/cneaf/electricity/esr/table5.html

Zhang, L. X.; Yang, Z. F.; Chen, B.; Chen, G. Q.; Zhang, Y. Q. (2009). Temporal and spatial variations of energy consumption in rural China. *Communications in nonlinear science and numerical simulation, 14*(11), 4022–4031.

future, will become an engine for innovation and ingenuity in the entire electricity sector.

The term *smart grid* subsumes many different aspects of electricity distribution and consumption. It can range from the mundane—like real-time electricity meters—to the truly revolutionary—like remote controls of air conditioners and other appliances that could allow utilities to steer customer demand via the grid.

Electricity use is not even throughout the year. Depending on climate, electricity use either peaks during the summer because of air conditioning or the winter because of heating, or both. Daily usage also varies. In homes, it generally peaks in the morning and evening, then declines at night. Utilities, however, must install enough capacity to meet peak demand, and much of that capacity goes unused most the time. (See the sidebar "What's the difference between a gigawatt and a gigawatt-hour?" on pages 4–5 for a further explanation of this often misunderstood topic.) Smart grid technology and a comprehensive view of the entire energy system can ensure that customers use electricity efficiently and that utilities need not overbuild to meet peak demand. The key is to keep the system as flexible and open as possible, while maintaining crucial oversight and central, long-term planning functions.

Regulators will need to help plan, fund, and even build smart grids. In the end, the regulatory task here is similar to that of pricing. Policy makers, rather than attempting to design the new grids and pick the winning technologies, should create a set of conditions and incentives that will motivate businesses and entrepreneurs to find innovative solutions to problems previously deemed intractable.

The Future

Renewable power is the energy of the future. It will move us from the high-carbon, low-efficiency world of today to a new world of low-carbon, high-efficiency economic development. Smart regulation can hasten this transition by capping global-warming pollution and creating a carbon market that provides incentives to build a new power system. The key to realizing this vision is for regulators to leave it to the market to pick and choose winning technologies and to maintain flexibility. The subsequent entries in this section provide a menu of possible options.

Gernot WAGNER
Environmental Defense Fund

See also in the *Berkshire Encyclopedia of Sustainability* Cap-and-Trade Legislation; Energy Efficiency; Energy Industries (*assorted articles*); Investment, CleanTech; Sustainable Value Creation; True Cost Economics

Further Reading

Dechezleprêtre, Antoine; Glachant, Matthieu; Hascic, Ivan; Johnstone, Nick; & Ménière, Yann. (2008). Invention and transfer of climate change mitigation technologies on a global scale: A study drawing on patent data. Retrieved November 5, 2009, from http://www.cerna.ensmp.fr/Documents/Invention_and_transfert_of_climate_mitigation_technologies_on_a_global_scale:_a_study_drawing_on_patent_data.pdf

International Energy Agency. (2009). *World energy outlook 2009.* Paris: Organisation for Economic Co-Operation and Development.

Krupp, Fred, & Horn, Miriam. (2008). *Earth, the sequel: The race to reinvent energy and stop global warming.* New York: W. W. Norton.

McKinsey & Company. (2009a). *Energy: A key to competitive advantage—New sources of growth and productivity.* Retrieved February 24, 2010, from http://www.mckinsey.com/clientservice/ccsi/pdf/Energy_competitive_advantage_in_Germany.pdf

McKinsey & Company. (2009b). *Unlocking energy efficiency in the US economy.* Retrieved November 5, 2009, from http://www.mckinsey.com/clientservice/ccsi/pdf/US_energy_efficiency_full_report.pdf

Stuebi, Richard. (2009). Money walks, fossil fuel talks. Retrieved March 3, 2010, from http://www.huffingtonpost.com/richard-stuebi/money-walks-fossil-fuel-t_b_300924.html

Sudarshan, Anant, & Sweeney, James. (2008). Deconstructing the "Rosenfeld Curve." Palo Alto, CA: Stanford University.

Tam, Cecilia, & Gielen, Dolf. (2007). *ETP 2008: Technology learning and deployment.* Retrieved November 6, 2009, from http://www.iea.org/textbase/work/2007/learning/Tam.pdf

United States Department of Energy (USDOE). (2009). *How the smart grid promotes a greener future.* Retrieved November 5, 2009, from http://www.oe.energy.gov/DocumentsandMedia/Environmentalgroups.pdf

Algae

Algae are a tremendously diverse group of organisms that have shaped the evolution of our planet. Commercial food products, goods used in food processing, and health products are obtained from algae, and they represent important markets for countries all over the world. With the continuous growth of human populations, the demands of these algal products have increased. As a result, new technologies and innovative approaches for algal research and utilization have been developed recently.

Algae (singular alga), from the Latin for "seaweed," are a highly diverse group of organisms generally composed of simple photoautotrophs (organisms that get energy from the sun). Algae occur mostly in aquatic habitats worldwide and in many different types of ecosystems and habitats. They range in size from microscopic forms (i.e., phytoplankton) to meters in length (giant kelps). Algae can also be found in terrestrial habitats, and they are especially abundant in tropical rain forests (Lopez-Bautista, Rindi, and Casamatta 2007). One of the most striking landmarks of our planet is the land plant, which evolved from an ancestral green alga millions of years ago. Several endosymbiotic events (where an organism becomes a part of another organism in a mutually beneficial relationship) during the long evolutionary history of our planet have resulted in a wide variety of present-day algae, and these algal groups are represented across the main branches of the tree of life. The chloroplast and the mitochondrion, two parts of most eukaryotic (with a true nucleus) cell machinery, are believed to have once been organisms that became components of cells via endosymbiotic events.

The major groups are classified into three algal phyla: Chlorophyta (*sensu lato*) or green algae, Rhodophyta or red algae, and the diverse Ochrophyta. Another group, the prokaryotic (without a nucleus) cyanobacteria or blue-green algae, are bacteria, which makes them very distinct from the rest of the algae. But cyanobacteria are studied by botanists as algae both for convenience and tradition. Algae have been a major changing force in our planet by providing oxygen to a primordial atmosphere devoid of oxygen and thus changing the face of our planet forever. Nowadays algae and their derived forms provide the oxygen we use for breathing and keeping our biosphere as it is. They are also responsible for more than 250 billion metric tons of sugars produced annually (Raven, Evert, and Eichhorn 2005). Both microscopic and macroscopic forms of algae have been used by humans in a variety of ways throughout history and in the present day; people first collected them from the wild and eventually developed large-scale farming and culture operations.

As Food for Humans

The oldest and simplest use of algae is as a food for humans. Marine macroalgae, colloquially called seaweeds, have been harvested for human consumption since at least 600 BCE in China (Guiry 2010). In the Western world, algae are generally regarded as a specialty or health food, but in Asia algae comprise an important fraction of people's diets; *kombu*, *wakame*, and *nori* are the Japanese names for three types of seaweed that are highly important on both cultural and economic levels (Guiry and Guiry 2010). Kombu refers to kelp, primarily *Saccharina japonica* (formerly in the genus *Laminaria*), a brown alga. This seaweed is the most widely cultivated marine organism in the world (Bixler and Porse 2010).

S. japonica has been harvested primarily in China, where it is called *haidai*, after new methods of farming were developed in the mid-1900s (McHugh 2003, 4). Wakame is a specific brown alga under the scientific name of *Undaria pinnatifida*, and it is harvested in a similar manner to that of *S. Japonica*. In China, both brown algal species are cultivated on hanging rafts. In France, however, the wakame's cultivation requires a different approach including partial cultivation in laboratory conditions. Due to a different temperature tolerance and local food preferences, *U. pinnatifida* is cultivated primarily in Japan and also in Korea, where the alga is called *miyeok* (McHugh 2003). Nori is the name given to various species of the red alga *Porphyra*, which is cultivated and consumed mainly in Japan, though *Porphyra* species in the past have been harvested in the wild in northern Europe as "laver" (Guiry and Guiry 2010). For centuries, nori was considered a luxury food because it was difficult to find in the wild or cultivate (McHugh 2003), but in 1949, Kathleen Drew (1949) discovered that *Porphyra* requires mollusk shells for part of its life cycle. She discovered that a cryptic filamentous form previously thought to be a completely different alga (called *Conchocelis*) was part of the nori life cycle. Since then, nori cultivation has grown, nori is more widely consumed, and it is harvested on an industrial scale. The economic impact of Drew's investigations was so significant that even to this day nori farmers celebrate the Drew Festival on April 14.

Many other edible algae are also harvested worldwide. These include, among other, dulse (*Palmaria palmata*) and Irish moss (*Chondrus crispus*) in Europe, *aonori* (*Enteromorpha* spp. and *Monostroma* spp.) in Japan, sea moss (*Gracilaria debilis)* in the Caribbean, sea grapes (*Caulerpa lentillifera)* in the Philippines, and *cochayuyo* (*Durvillaea antarctica*) in Chile. Seaweeds in general are a good source of vitamins and minerals and sometimes protein, though the proportions vary between species and harvest times (Noda 1993; Yamanaka and Akiyama 1993).

Microalgae are much less widely cultivated than macroalgae, but several cultures have developed ways to use microalgae as food. The most notable microalgal food source is a filamentous cyanobacterium widely called spirulina; this name comes from the genus *Spirulina*, though the algae in question have since been renamed *Arthrospira platensis* and *Arthrospira maxima* (Kómarek and Hauer 2010). Spirulina has served as a food (*tecuitlatl*) for the precolonial Aztecs (Ortega 1972) as well as several present-day African cultures (Chamorro et al. 1996). Although *Arthrospira* is a microorganism, it often grows in mats that can be skimmed off the surface of water in a wide variety of environments, requiring little in terms of resources. Along with the fact that spirulina is a very good source of protein and other nutrients, the ease of cultivation of spirulina has made it appealing for use in developing countries in which hunger is a widespread problem (IIMSAM 2006). Spirulina and other cyanobacteria such as *Aphanizomenon flos-aquae* (Pugh et al. 2001) are also used as nutritional supplements or food additives, with various health benefits cited as a result of their consumption. In addition to any vitamins present, examples of these purported health benefits include immunostimulatory (Pugh et al. 2001) and antioxidant properties (Wu et al. 2005). Conversely, certain cyanobacteria have been known to produce toxins affecting humans and economically important fisheries. These toxins include beta-methylamino-L-alanine (BMAA), saxitoxin, and microcystin, raising the question of whether the consumption of cyanobacteria is safe (Gilroy et al. 2000). Biomagnification of BMAA in particular has been observed in studies of the Chamorro people of Guam, where the incidence of neurodegenerative disease is fifty to one hundred times the incidence elsewhere (Cox, Banack, and Murch 2003). But recent investigation challenges a correlation between the neurodegenerative disease and the BMAA produced by cyanobacteria (Snyder et al. 2009). Some cyanobacterial toxins have been proven useful in pharmacology, such as tolytoxin produced by the cyanobacterium *Tolypothrix* (Graham, Wilcox, and Graham 2009). Besides cyanobacteria, some green microalgae are used in industries; the alga *Dunaliella* is harvested to synthesize beta-carotene (Mojaat et al. 2007) to be used in supplements, and *Chlorella* is sold as a supplement similar to spirulina (Pugh et al. 2001).

Phycocolloids: Gels from Algae

Cell walls of red and brown seaweeds contain thick polysaccharide gels (substances made of carbohydrates that give the cell structure) that have become useful in food processing and in other industries. These gels, called phycocolloids, are extracted from different types of algae to manufacture three primary thickening agents: alginate from brown algae, and agar and carrageenan from red algae (McHugh 2003). An entire industry surrounding mass production of these gels has developed and continues to grow: in 2009, total sales of phycocolloids accounted for 86,100 metric tons (MT), worth over US$1 billion, compared to 1999, with 72,500 MT worth $644 million (Bixler and Porse 2010). All three kinds of phycocolloids are used to help emulsify, bind, thicken, and otherwise improve the texture of foods, such as ice cream and mayonnaise, as well as other products like toothpaste and paint (Guiry and Guiry 2010).

Alginates are extracted from brown algae, also known as kelps. They were discovered in the late nineteenth century in Scotland, and an alginate industry developed in the 1940s. Genera harvested to produce alginate (sometimes referred to as algin) have changed dramatically since 2000. In 1999, *Macrocystis* from the Americas and *Ascophyllum* from Europe were the most-harvested seaweeds for alginate, but the alginates extracted from these two genera are low in guluronic acid, which results in a less rigid gel (Bixler and Porse 2010). In addition, extracts of *Ascophyllum* are very dark, resulting in a product that must be bleached strongly (McHugh 2003). In 2009 the most important alginate seaweeds had shifted to *Laminaria* from Europe and Asia and *Lessonia* from South America, these two genera accounting for a total of 81 percent of seaweed harvests by weight. Combined, *Macrocystis* and *Ascophyllum* accounted for 58 percent of harvests in 1999, but by 2009 the amount had dropped to only 8 percent (Bixler and Porse 2010). *Laminaria japonica*, also called *Saccharina japonica* (Guiry and Guiry 2010), is easily the world's largest maricultural crop, grown in huge quantities in China, though not all of it is used for alginate. To extract alginate, the seaweed is pulverized and heated with an alkali solution, then diluted and filtered to remove seaweed residue. The solution then is chemically treated to eventually form a paste of sodium alginate, which can be dried and sold (McHugh 2003); the average price for alginate is about US$12 per kilogram (Bixler and Porse 2010). Alginate sales amounted to 26,500 MT and US$318 million in 2009 (Bixler and Porse 2010), representing about 30 percent of total phycocolloid sales value. In addition to food-related applications, alginates are also used in the textile pigmentation (McHugh 2003) and papermaking industries (Bixler and Porse 2010).

Agar is produced almost exclusively from two genera of red algae, *Gracilaria* and *Gelidium*. Agar was first discovered in Japan before the seventeenth century. The agar extracted from *Gracilaria* is used in foods, but *Gelidium* produces a high-quality agar that is crucial in bacteriology as the basis for culture media. To extract agar, the seaweed is heated in water for several hours, dissolving the agar in the water. After filtering, a jelly is produced with around 1 percent agar, which is then concentrated and dried (McHugh 2003). In 2009, 80 percent of manufactured agar was agar powder from *Gracilaria* (Bixler and Porse 2010), and most agar is produced in Asia, followed by the Americas. Of the three phycocolloid industries, the agar industry has the smallest sales volume—only about 15 percent of the total, having increased from 7,500 MT in 1999 to 9,600 MT in 2009—and the highest average price, worth about US$18 per kilogram (Bixler and Porse 2010). Agarose, the isolated gelling component of agar, is highly important in biotechnology and can reach prices upwards of US$5,000 per kilogram (Guiry and Guiry 2010). The market for agarose is small, consisting of less than 10 percent of the overall market for agar (Bixler and Porse 2010), and agarose producers usually buy agar rather than processing seaweed for their starting material (McHugh 2003). Besides its scientific applications, agar can be used as a vegetarian alternative to gelatin or in baking.

Of the various carrageenans, three are commercially useful based on their chemical properties: lambda, kappa, and iota (McHugh 2003). Kappa carrageenan forms firm gels, while iota forms elastic gels; lambda carrageenan does not form a gel but thickens solutions. Carrageenans are extracted from different red algal species, and the chemical composition of the extracts varies based on the algal species used. For example, *Chondrus crispus* produces a mixture of kappa and lambda carrageenans, *Kappaphycus alvarezii* produces mainly kappa, and *Eucheuma denticulatum* produces mainly iota (McHugh 2003). The carrageenan industry has grown from a sales volume of 42,000 MT to 50,000 MT since 2000. In addition, the average price of carrageenans has risen from US$7 per kilogram to US$10.50 per kilogram, resulting in the total sales value of carrageenans increasing from US$291 million to US$527 million (Bixler and Porse 2010). The majority of seaweeds (80 percent) harvested for carrageenan belong to the species *Kappaphycus alvarezii* (commonly harvested in the Philippines), while *Chondrus crispus*, although historically important as a carrageenan source, only represents 2 percent of 2009 harvests (Bixler and Porse 2010). Carrageenan is used in processed meats to increase product yields or replace fat, in dairy products as a stabilizer, and in canned pet foods to prevent separation of fat. The Food and Agriculture Organization/World Health Organization (FAO/WHO) Expert Committee on Food Additives determined that adding carrageenan is not a health hazard and is acceptable for daily intake (Guiry and Guiry 2010).

Algae as Environmental Indicators

Algae have proven useful as environmental bioindicators and bioassays. *Bioindicator* refers to the practice of directly examining an environmental sample, while *bioassay* refers to adding nutrients, like nitrogen and phosphorus, to different samples and examining any effects. Based on a variety of environmental factors, the types of algae present in a water sample, or the number of algal cells of some species in a sample, can be a source of information reflecting the chemical content of the environment (Graham, Wilcox, and Graham 2009). While it is possible to analyze water samples using chemical means, these methods can over- or underestimate the biologically available amount of the chemical. Round (1981) discussed many different species that can be used to measure different properties of a water sample, including

Selenastrum capricornutum. This species is a unicellular green alga that is the most widely used alga in bioassays (Graham, Wilcox, and Graham 2009); more recently it has been established that *Pseudokirchneriella subcapitata* is the correct name for *Selenastrum capricornutum* (Guiry and Guiry 2010). Many other studies use multimetric analyses of the phytoplankton (i.e. microalgae), based on abundance of different groups of algae or the presence of certain algal molecules, like chlorophyll *a* (Lacouture et al. 2006). With recent advances in genomics, the concept of an "ecogenomic sensor" has arisen, which is a device that can be placed in an environment and continuously take samples, analyzing the DNA present to detect organisms or genes over time (Scholin 2010). New technologies may soon be used regularly to remotely detect certain algae that have been shown to produce harmful algal blooms (HAB), toxins, or other deleterious environmental effects (Scholin et al. 2009).

Algae in Biofuel Production

Biofuels are a renewable resource and an alternative to the use of fossil fuels. Currently the primary biofuel industry is based on the production of ethanol from plant biomass, including corn (maize) and sugarcane (Demirbas and Demirbas 2010). Bioethanol is often mixed with gasoline in various concentrations to reduce the need for fossil fuels; in the United States, gasoline is mandated to contain up to 10 percent ethanol (RFA 2010), while in Brazil, 25 percent ethanol or higher is standard. There are several challenges with the production of biofuels from these crops, however, including competition with human food production (i.e., "food vs. fuel"), destruction of natural environments, and differences in greenhouse gas emissions between bioethanol and regular gasoline (UNEP 2009). The use of algae as a source for biofuels could provide a solution for some of these problems (Clarens et al. 2010). The algae used to produce biofuels are not food crops, and they would therefore not be involved in the food vs. fuel debate. Also, algal bioreactors can be built on unused land unsuitable for farming (i.e., deserts), which would not lead to deforestation (Greenwell et al. 2010). Algal bioreactors require large quantities of nitrogen, phosphorus, water, and carbon dioxide added to the system in order to be efficient, but their carbon footprint can be greatly reduced when built in conjunction with a wastewater treatment plant (Greenwell et al. 2010), a residual source of nitrogen and phosphorus. Many different processes can be used to make biofuels from microalgae, including the production of biodiesel, a fuel created from naturally occurring lipids that can completely replace petroleum-based diesel fuel used in transportation. Oil is reacted with methanol to produce methyl esters, which are then used as fuel, and glycerin is produced as a byproduct (Chisti 2007). The oils currently used to produce biodiesel can be obtained from several different sources, including large-scale rapeseed farms, waste vegetable oil from food factories and restaurants, and algal bioreactors. Many different kinds of microalgae produce large amounts of lipids naturally; the algae are then pulverized, releasing the oil. In this process, algae yields a high percentage of oil compared to oil crops like rapeseed or soybeans. Combined with the high volume of algae production, as well as the advantages of algal bioreactors, algal biodiesel has great potential as a renewable, carbon-neutral energy source. Due to the limited amount of research on the subject, however, the economic feasibility of mass production of algal biodiesel is uncertain (Scott et al. 2010). Other applications of algae in biofuel production involve hydrogen gas, which can be used as a fuel itself (Hankamer et al. 2007; Kruse and Hankamer 2010) or mixed with carbon monoxide to produce syngas, which is used to produce diesel fuel by the Fischer-Tropsch synthesis (Demirbas and Demirbas 2010).

New Trends and Implications

Algal resources will play a significant role in our planet's welfare during the next decades. Previous and current models of algal utilization have made use of *exterior* conditions in order to improve algal crops and/or harvesting. The new models, however, are instead altering *internal* conditions to permanently modify the algal DNA. Algal transgenics, or the transformation of algal cells using genetic engineering techniques, is a new and rapidly evolving biotechnological field. With the advancements of algal genomes and innovating biotechniques it has been possible to "synthesize" algal cells with desirable traits. These "synthetic organisms" are a reality, and new patents have been already filed (Walker, Collet, and Purton 2005). Depending on the desired commercial product (Hallmann 2007; Cardozo et al. 2007), efforts are being made to create algal cells to function as cell-factories (León-Bañares et al. 2004). Several companies are applying transgenics technologies to improve algal cells and their commercial products (Waltz 2009) and "next generation biofuels" (SGI 2010). The new field of algal transgenics is not without difficulties: biotechnology and genetic engineering problems as well as public awareness and biosafety concerns need to be addressed (Hallmann 2007). But if the promise of "the green gold" becomes a reality in the next years, algae will undoubtedly alleviate many concerns for the near and distant future of the human race.

Juan LOPEZ-BAUTISTA and Michael S. DePRIEST
The University of Alabama

The authors greatly appreciate facilities at the University of Alabama and funds from the National Science Foundation through their Tree of Life Programs to JL-B (DEB 0937978, DEB 1027012, and DEB 1036495).

See also in the *Berkshire Encyclopedia of Sustainability* Agriculture (*several articles*); Bioenergy and Biofuels; Food (*several articles*); Hydrogen Fuel

Further Reading

Bixler, Harris J., & Porse, Hans. (2010, May 22). A decade of change in the seaweed hydrocolloids industry. *Journal of Applied Phycology: Online First*. Retrieved September 29, 2010, from http://www.algaebase.org/pdf/AC100CF011cce16156PIqW9AFCE2/Bixler_Porse.pdf

Cardozo, K. H. M., et al. (2007). Metabolites from algae with economical impact. *Comparative Biochemistry and Physiology*, *C*(146), 60–78.

Chamorro, Germán; Salazar, María; Favila, Luis; & Bourges, Héctor. (1996). Farmacología y toxicología del alga Spirulina [Pharmacology and toxicology of Spirulina alga]. *Revista de Investigación Clinica*, *48*(5), 389–399.

Chisti, Yusuf. (2007). Biodiesel from microalgae. *Biotechnology Advances*, *25*(3), 294–306.

Clarens, Andres F.; Resurreccion, Eleazer P.; White, Mark A.; & Colosi, Lisa M. (2010). Environmental life cycle comparison of algae to other bioenergy feedstocks. *Environmental Science & Technology*, *44*(5), 1813–1819.

Cox, Paul A.; Banack, Sandra Anne; & Murch, Susan J. (2003). Biomagnification of cyanobacterial neurotoxins and neurodegenerative disease among the Chamorro people of Guam. *Proceedings of the National Academy of Sciences of the United States of America*, *100*, 13380–13383.

Demirbas, Ayhan, & Demirbas, M. Fatih. (2010). *Algae energy: Algae as a new source of biodiesel*. London: Springer.

Drew, Kathleen M. (1949). Conchocelis-phase in the life-history of Porphyra umbilicalis (L.) Kütz. *Nature*, *164*, 748–749.

Gilroy, Duncan J.; Kauffman, Kenneth W.; Hall, Ronald A.; Huang, Xuan; & Chu, Fun S. (2000).

Assessing potential health risks from microcystin toxins in blue-green algae dietary supplements. *Environmental Health Perspectives*, *108*(5), 435–439.

Graham, Linda E.; Wilcox, Lee W.; Graham, James M. (2009). *Algae*. San Francisco: Pearson.

Greenwell, H. C.; Laurens, L. M. L.; Shields, R. J.; Lovitt, R. W.; & Flynn, K. J. (2010). Placing microalgae on the biofuels priority list: A review of the technological challenges. *Journal of the Royal Society Interface*, *7*(46), 703–726.

Guiry M. D., & Guiry, G. M. (2010). AlgaeBase. [World-wide electronic publication, National University of Ireland, Galway.] Retrieved September 13, 2010, from http://www.algaebase.org

Guiry, Michael. (2010). Seaweed site. Retrieved September 13, 2010, from http://seaweed.ucg.ie/index.html

Hallmann, Armin. (2007). Algal transgenics and biotechnology. *Transgenic Plant Journal*, *1*(1), 81–98.

Hankamer, B., et al. (2007). Photosynthetic biomass and H2 production by green algae: From bioengineering to bioreactor scale-up. *Physiologia Plantarum*, *131*(1), 10–21.

Intergovernmental Institution for the Use of Micro-Algae Spirulina Against Malnutrition (IIMSAM). (2006). Benefits of spirulina. Retrieved August 17, 2010, from http://www.iimsam.org/benefits.php

Komárek, Jiří, & Hauer, Tomáš. (2010). Arthrospira. Retrieved August 17, 2010, from http://www.cyanodb.cz/Arthrospira

Kruse, Olaf, & Hankamer, Ben. (2010). Microalgal hydrogen production. *Current Opinion in Biotechnology*, *21*(3), 238–243.

Lacouture, Richard V.; Johnson, Jacqueline M.; Buchanan, Claire; & Marshall, Harold G. (2006). Phytoplankton index of biotic integrity for Chesapeake Bay and its tidal tributaries. *Estuaries and Coasts*, *29*(4), 598–616.

Le ón-Bañares, Rosa; Gonz ález-Ballester, David; Galván, Aurora; & Fernández, Emilio. (2004). Transgenic microalgae as green cell-factories. *TRENDS in Biotechnology*, *22*(1), 45–52.

Lopez-Bautista, Juan M.; Rindi, Fabio; & Casamatta, Dale. (2007). The systematics of subaerial algae. In Joseph Seckbach (Ed.), *Cellular origin, life in extreme habitats and astrobiology: Volume 11. Extremophilic algae, cyanobacteria and non-photosynthetic protists: From prokaryotes to astrobiology* (pp. 599–617). Dordrecht, The Netherlands: Springer.

McHugh Dennis J. (2003). *A guide to the seaweed industry* (FAO Fisheries Technical Paper 44). Rome: Food and Agriculture Organization of the United Nations.

Mojaat, M.; Foucault, A.; Pruvost, J.; & Legrand, J. (2007). Optimal selection of organic solvents for biocompatible extraction of beta-carotene from Dunaliella salina. *Journal of Biotechnology*, *133*(4), 433–441. Noda, Hiroyuki. (1993). Health benefits and nutritional properties of nori. *Journal of Applied Phycology*, *5*(2), 255–258.

Ortega, Martha Ma. (1972). Study of the edible algae of the Valley of Mexico. *Botanica Marina*, *15*, 162–166.

Phycological Society of America. (2010). Homepage. Retrieved September 13, 2010, from http://www.psaalgae.org/

Pugh, Nirmal; Ross, Samir A.; ElSohly, Hala N.; ElSohly, Mahmoud A.; & Pasco, David S. (2001). Isolation of three weight polysaccharide preparations with potent immunostimulatory activity from Spirulina platensis, Aphanizomenon flos-aquae and Chlorella pyrenoidosa. *Planta Medica*, *67*, 737–742.

Raven, Peter H.; Evert, Ray F.; & Eichhorn, Susan E. (2005). *Biology of Plants* (7th ed.). New York: W. H. Freeman.

Renewable Fuels Association (RFA). (2010). 2010 annual RFA ethanol industry outlook: Climate of opportunity. Retrieved September 13, 2010, from http://www.ethanolrfa.org/page/-/objects/pdf/outlook/RFAoutlook2010_fin.pdf?nocdn=1

Round, F. E. (1981). *The ecology of algae*. Cambridge, UK: Cambridge University Press.

Scholin, Christopher, et al. (2009). Remote detection of marine microbes, small invertebrates, harmful algae, and biotoxins using the Environmental Sample Processor (ESP). *Oceanography*, *22*(2), 158–161.

Scholin, C. A. (2010). What are "ecogenomic sensors?" A review and thoughts for the future. *Ocean Science*, *6*, 51–60.

Scott, Stuart A., et al. (2010). Biodiesel from algae: challenges and prospects. *Current Opinion in Biotechnology*, *21*(3), 277–286.

Snyder, L. R., et al. (2009). Lack of cerebral BMAA in human cerebral cortex. *Neurology*, *72*, 1360–1361.

Synthetic Genomics Inc. (SGI). (2010). SGI corporate overview. Retrieved September 29, 2010, from http://www.syntheticgenomics.com/images/SGI-overview.pdf

United Nations Environment Programme (UNEP). (2009). *Towards sustainable production and use of resources: Assessing biofuels*. Nairobi, Kenya: United Nations Environment Programme.

Walker, Tara L.; Collet, Chris; & Purton, Saul. (2005). Review: Algal transgenic in the genome era. *Journal of Phycology*, *41*, 1077–1093.

Waltz, Emily. (2009). Biotech's green gold? *Nature Biotechnology*, *27*, 15–18.

Wu, Li-chen; Ho, Ja-an Annie; Shieh, Ming-Chen; & Lu, In-Wei. (2005). Antioxidant and antiproliferative activities of spirulina and chlorella water extracts. *Journal of Agricultural and Food Chemistry*, *53*(10), 4207–4212.

Yamanaka, Ryoichi, & Akiyama, Kazuo. (1993). Cultivation and utilization of Undaria pinnatifida (wakame) as food. *Journal of Applied Phycology*, *5*(2), 249–253.

Aluminum

An increasing supply of aluminum is essential to continued global economic development. Production is unlikely to be constrained by raw materials until the second half of the twenty-first century but may eventually require alternative ores as reserves of bauxite are consumed. Increased recycling and continuous reductions in energy usage, greenhouse gas generation, and toxic emissions are priorities for the long-term sustainability of the production process.

Major periods in human history have been named according to the material that defined them—such as the Stone, Iron, and Bronze Ages. The influence of aluminum since the early twentieth century has been transformative to the extent that we are arguably now living in the "Aluminum Age." Aluminum is present in a wide range of minerals, primarily oxides and hydroxides. As such, it comprises about 8 percent of the Earth's crust, in which it is the third most abundant element after oxygen and silicon.

Aluminum (Al) was unknown in metallic form until 1808, and large-scale industrial production was not possible until the 1880s. The subsequent rapid growth of the industry was enabled by extraordinary innovation and technological development. Thanks to aluminum's unique combination of natural abundance, light weight, high electrical and heat conductivity, corrosion resistance, and recyclability, it has become a commonplace material in all aspects of modern life. For example, aluminum is the basis of most of the lightweight alloys in use today, without which manned flight, particularly large-scale air travel and space exploration, would have been difficult if not impossible to achieve.

Aluminum is a highly reactive metal, but in air it is covered with a tenacious oxide skin that protects it. Because of this, aluminum is, in practical terms, generally not reactive and is far more corrosion-resistant than iron. It is this remarkable property of surface passivation that underpins the utility of aluminum and its alloys. In addition, aluminum is light (being only 30 percent the density of steel), is an excellent conductor of heat and electricity, and is essentially nontoxic to humans (although there is ongoing debate about a possible role in relation to Alzheimer's disease).

Production

The first commercial production of aluminum, by reduction of aluminum oxide with sodium, was in France in 1854. Kilogram amounts of aluminum could be made this way, but with great difficulty and expense, so aluminum originally was a rare and precious metal. True industrial production awaited the technological breakthrough based on electrical smelting of alumina (aluminum oxide) in a bath of molten cryolite (sodium aluminum fluoride), which was discovered in 1888 simultaneously by Charles Hall in the United States and Paul Héroult in France, and the process for the production of the key feedstock, alumina, from bauxite by Austrian scientist Karl Josef Bayer in 1889. These new production techniques resulted in a dramatic drop in the price of aluminum and made possible its use as a commodity metal with many and varied high-volume uses (Wallace 1937).

From its inception in the 1890s, aluminum production has increased exponentially (Roskill 2008). The correlation with the US gross domestic product

Figure 1. Growth of Global Aluminum Production Correlated with the Growth of the US Economy

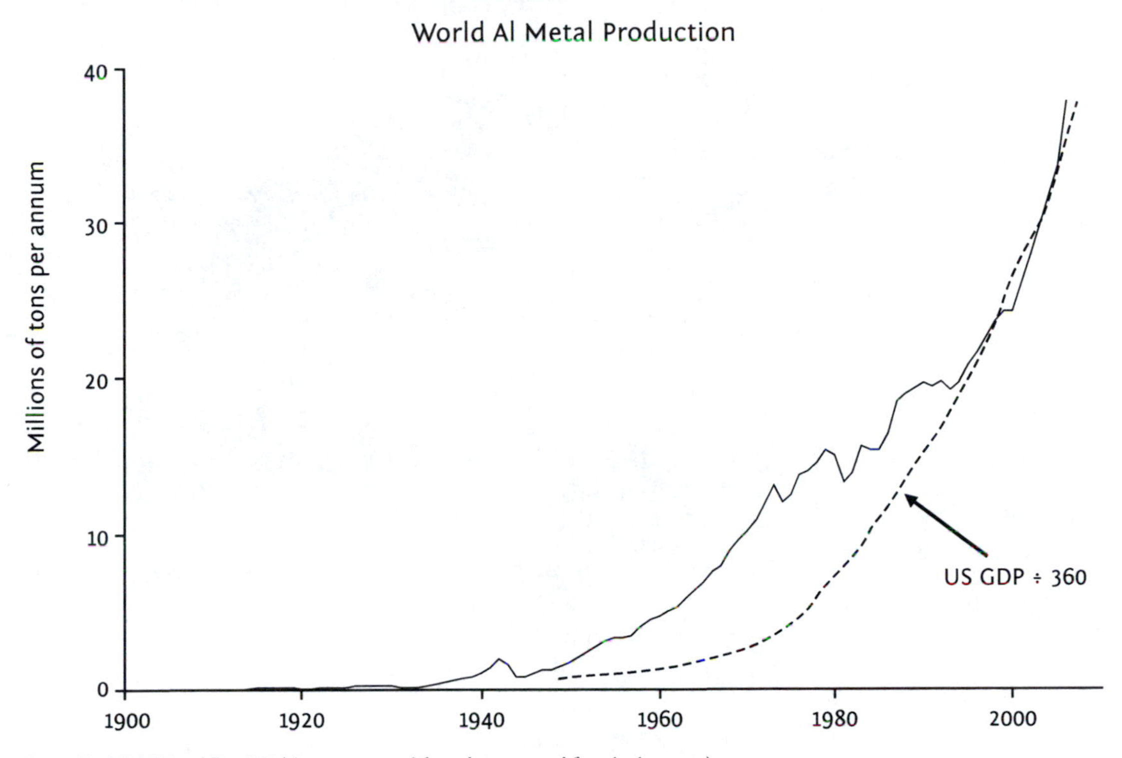

Source: Roskill 2008 and Data360 (chart constructed from data extracted from both sources).

Aluminum production has increased exponentially since the 1890s. The correlation with the United States' gross domestic product (GPD) indicates that aluminum production and overall economic activity are intimately connected.

(Data360 2011) indicates that aluminum production goes hand-in-hand with overall economic activity, as shown in figure 1.

Uses

Aluminum found a large and ever-growing range of uses in the twentieth century and was the basis of a number of advances that had profound influences on the development of technology, and through it, on the nature of human society. It was fundamental to manned flight: the engine that powered the Wright Brothers' aircraft in 1903, a mere fifteen years after the Hall/Héroult patents, featured an aluminum alloy engine block to achieve the required power-to-weight ratio. It is unlikely that mass air travel would have been possible without the availability of aluminum alloys for airframes, cladding, and engine parts. Aluminum is ubiquitous in the modern world. Some other key uses include: food and beverage containers and cooking utensils; high performance sports equipment such as lightweight bicycles, baseball bats, tennis rackets, and golf clubs; window frames, structural supports, and cladding in buildings; alloy engine blocks, wheels, car frames, and cladding for vehicles; and high-reflectance mirror coatings. Its inherent reactivity also makes it useful as a pyrophoric agent (sparking for ignition) in fireworks, chemical welding, and rocketry—the brilliant white plume emitted in the launch of space rockets is alumina formed from the reaction of aluminum powder with oxygen. The main categories of use in the United States are summarized for 2008 in figure 2 on the following page. The first such survey was completed thirteen years earlier (Margolis 1997); the data shows that usage patters have been quite stable over that period.

Figure 2. Use of Aluminum in the United States

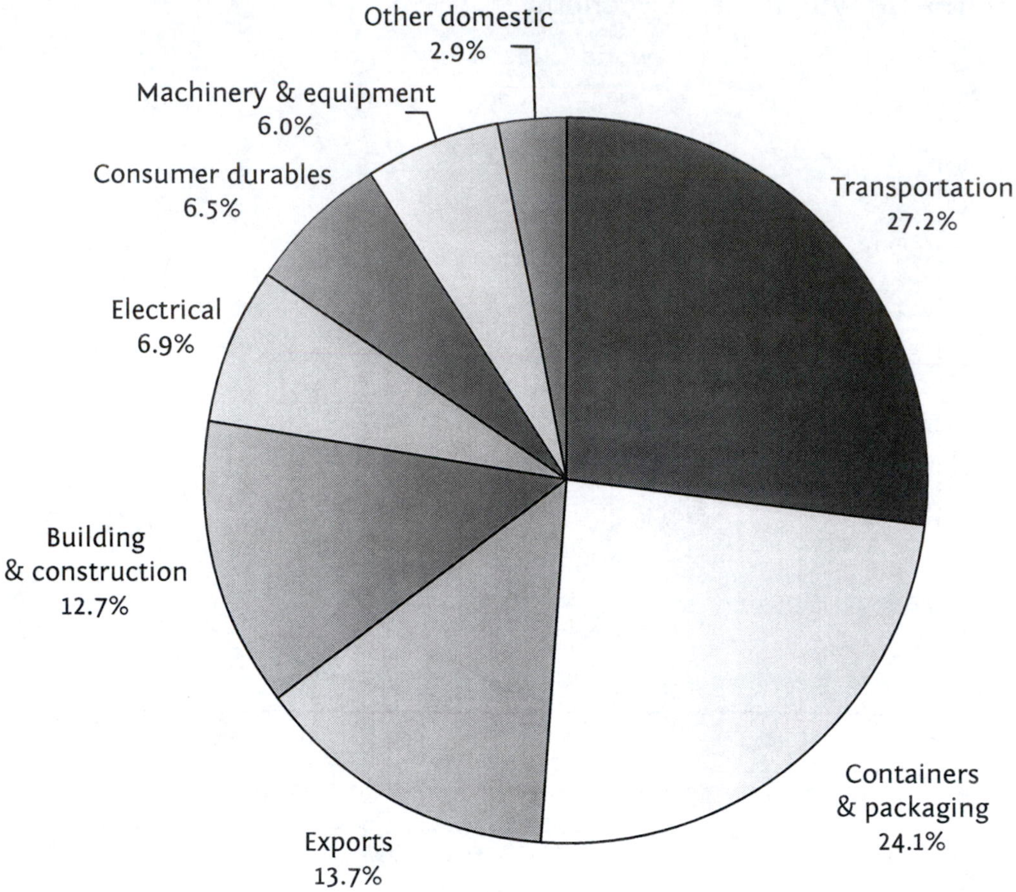

Other domestic 2.9%
Machinery & equipment 6.0%
Consumer durables 6.5%
Electrical 6.9%
Building & construction 12.7%
Exports 13.7%
Containers & packaging 24.1%
Transportation 27.2%

Source: Adams 2009.

Sustainability: Strengths, Issues, and Opportunities

The aluminum industry is aware of the need to engage in sustainable production practices at all stages, from extraction to processing. While the production process is carbon dioxide (CO_2) intensive, aluminum is also easily recycled.

Raw Material Supply

About 87 percent of the bauxite mined annually is used directly for the production of aluminum (Roskill 2008). The 2008 aluminum production of 38 million tons required about 150 million tons of bauxite. With the reserve base of bauxite estimated to be 32,300 million tons (Roskill 2008), bauxite is unlikely to be limiting until the second half of the twenty-first century, although continuous supply for increasing production will require the continuous development and expansion of new and existing deposits. Beyond bauxite, there are virtually unlimited potential resources of aluminum worldwide in other minerals, especially clays, which could in principle be exploited.

The other main raw materials required for aluminum production are caustic soda and carbon. Most of the caustic soda used today by the aluminum industry is a by-product of the chlor-alkali industry; this linkage is a leading example of industrial synergy. If this source becomes constrained, the alternative is to react calcine natural soda ash (sodium carbonate) with quicklime (calcium oxide).

Carbon is used to line the pots and for the anodes used in electrolysis of alumina to aluminum. The anodes are consumed in the process, according to the equation:

$$2Al_2O_3 + 3C \rightarrow 4Al + 3CO_2$$

This requires theoretically 0.33 tons and practically about 0.5 tons of carbon for each ton of aluminum (Alcoa n.d.), and an equivalent amount of CO_2 is produced. Anodes are made from pitch, which is not likely to become supply constrained.

Bauxite Mining and Residue Storage

The best bauxite occurs in tropical and subtropical regions, often under forests and mostly in shallow deposits amenable to open-cut mining. Australia is the largest producer of bauxite, accounting for 30 percent of global production in 2007, followed (in order) by Brazil, China, Guinea, India, Indonesia, and Jamaica, which together account for 50 percent. The largest reserve is in Guinea (Roskill 2008). The bauxite mining industry has become a world leader in mine rehabilitation and has won a number of environmental awards (Alcoa 2011). In 2006, about 30 square kilometers (equivalent to about half the area of Manhattan Island) was disturbed for mining, and a similar area was rehabilitated. Achievement of this balance is key to the sustainability of global bauxite mining (IAI 2009).

Management of bauxite residue, which is the mixture of caustic mud and sand that remains after aluminum has been extracted, has been identified as a key sustainability issue for the industry (IAI 2006, 54). Current best practice discourages disposal practices such as marine dumping and is focused on secure land-based storage (Power, Gräfe, and Klauber 2011, 33–45). The industry's 120-year unblemished safety record was tarnished in 2010 by the failure of a dam wall in Hungary, which caused several deaths, the destruction of property, and contamination of land and waterways (Szandelszky and Gorondi 2010). While such disasters can be avoided through sound engineering and management practices, the longer-term vision is for reintegration of the residue with the environment (Gräfe, Power, and Klauber 2011, 60–79) and/or using it for the manufacture of other products (Klauber, Gräfe, and Power 2011, 11–32).

Energy Use and Greenhouse Gases

The only practical way to produce aluminum is by electrolyzing alumina. This requires a great deal of electricity, partly because of the very high chemical stability of alumina and also because the electrolysis cell is maintained above 900°C by electrical heating, resulting in a requirement of about 115,000 megajoules per ton of aluminum at an energy efficiency of about 50 percent (Margolis 1997). Low-cost energy is essential. Many smelters are therefore near hydroelectric generators, which also reduce the greenhouse gas (GHG) intensity of the process. Significant amounts of GHG are nevertheless produced by the reaction of the carbon anodes, which results in 1.22 tons of CO_2 per ton of aluminum. Attempts to reduce this by the use of alternative (inert) anode materials have so far been unsuccessful (Cheatham et al. 1998). The overall CO_2 intensity is much larger when fossil fuel energy sources are used, up to 14 tons of CO_2 per ton of aluminum if the energy is from low rank coals (Turton 2002). Much of the reporting of energy efficiency and GHG production by the industry relates to the smelter itself and allocates the contribution of electricity generation to the third-party supplier; the full picture, and effective mitigation, relies on the inclusion of the generation aspect to the production of the metal in the calculation of its overall GHG contribution.

Emissions and Safety

Apart from CO_2, the main emissions to air from aluminum smelters are fluorides and fluorocarbons. Fluorides have been implicated in direct human health effects (e.g., in an occupational disease known as "potroom asthma," which has been known to afflict workers in smelters) and indirect effects through entry to the food chain through cow's milk and vegetables

Figure 3. Life Cycle of Aluminum

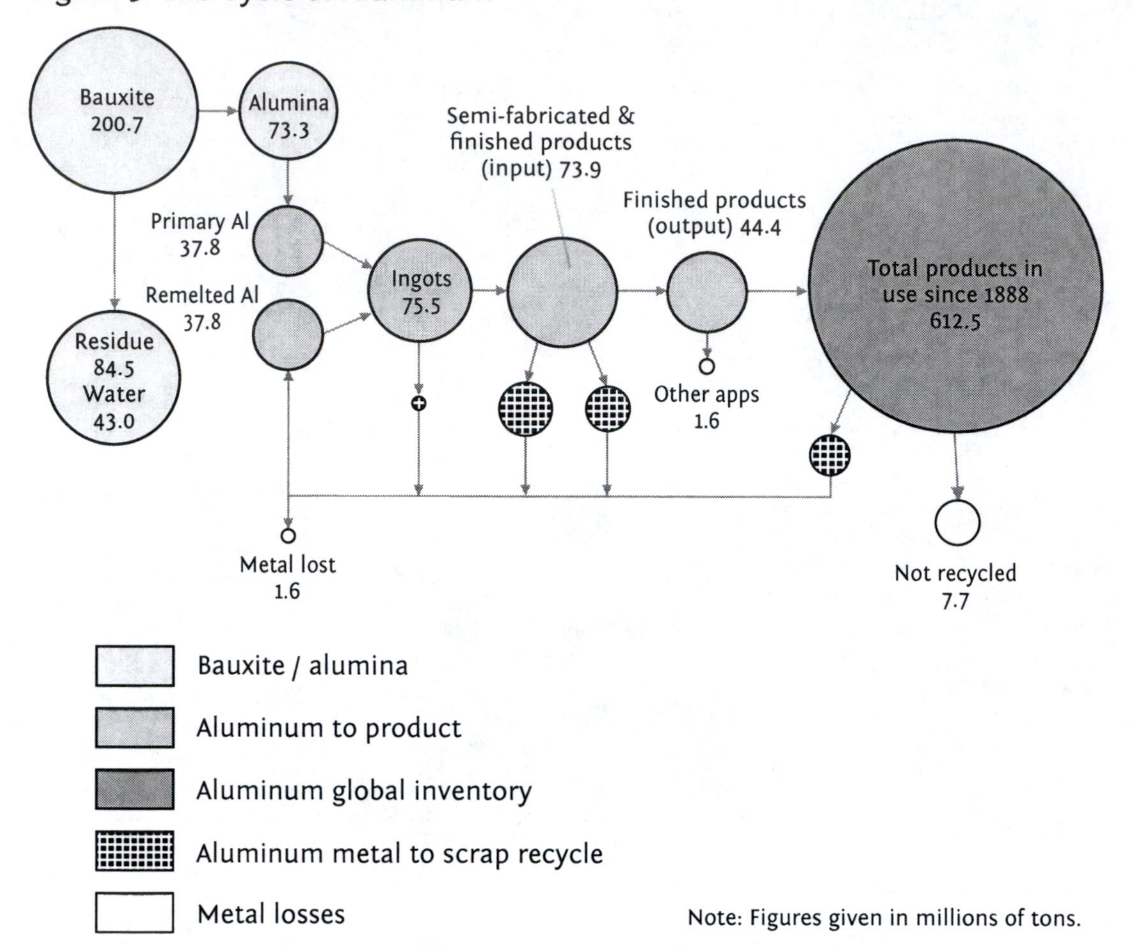

Source: IAI 2009, redrawn from 2008 chart.

downwind of smelters. Such effects have been effectively eliminated through modern cell design and efficient scrubbing of effluent gases.

Emissions of fluorocarbons, which are powerful GHGs, occur through the reaction of the anodes with the electrolyte bath (cryolite, sodium aluminum fluoride), mainly during transient anode effects. Reduction of anode effects through improvements in cell design and operation has led to major reductions in fluorocarbon emissions.

Recycling

Aluminum recycling is driven by its energy advantage: it needs only between 5 and 8 percent of the energy of primary production (AAC 2009; Plunkert 2006). Currently, about 50 percent of aluminum comes from recycled metal, and the industry is committed to continuously improving that figure (IAI 2011). The main flows in the aluminum life cycle are summarized in the flow chart shown in figure 3, above, which is updated annually by the International Aluminum Association. Note also that over three-quarters of the total amount of aluminum ever produced is still in use (Martchek 2006, 34–37).

Future Prospects

Primary aluminum production can be expected to continue to increase, driven by global economic development, and should not be constrained by raw materials supply into the second half of the twenty-first century.

The industry has a good record of recycling, which it is committed to improving. While aluminum must compete with other materials in specific applications, its overall versatility should mean that this competition is not a strategic threat to the industry. A key challenge for the long-term sustainability of the industry is to reduce overall CO_2 emissions intensity (in which it is essential to account for the primary electricity supply in the calculation); this need is being addressed in the short term through efficiency improvements, but ultimate solutions rest with low CO_2 energy sources and inert anodes.

Greg POWER
Arriba Consulting Pty Ltd

See also in the *Berkshire Encyclopedia of Sustainability* Chromium; Coltan; Copper; Electronics—Raw Materials; Gold; Heavy Metals; Iron Ore; Lead; Lithium; Manganese; Minerals Scarcity; Mining—Metals; Nickel; Platinum Group Elements; Rare Earth Elements; Recycling; Silver; Thorium; Tin; Titanium; Uranium

Further Reading

Adams, Nick. (2009, September). Facts at a glance—2008. The Aluminum Association, Inc., Industry Statistics. Retrieved June 1, 2011, from http://www.aluminum.org/Content/NavigationMenu/NewsStatistics/StatisticsReports/FactsAtAGlance/factsataglance.pdf

Alcoa. (n.d.). Aluminum smelting. Retrieved May 19, 2011, from http://www.alcoa.com/global/en/about_alcoa/pdf/Smeltingpaper.pdf

Alcoa. (2011). Awards. Retrieved May 19, 2011, from http://alcoa.com/australia/en/info_page/mining_awards.asp

Australian Aluminium Council (AAC). (2009). Sustainability report 2009. Retrieved May 19, 2011, from http://aluminium.org.au/sustainability-report

Cheatham, Bob, et al. (1998). Inert anode roadmap. Energetics Inc. Retrieved June 14, 2011, from http://www1.eere.energy.gov/industry/aluminum/pdfs/inertroad.pdf

Data360. (2011). GDP United States. Retrieved May 19, 2011, from http://data360.org/dsg.aspx?Data_Set_Group_Id=230

Gräfe, Markus; Power, Greg; & Klauber, Craig. (2011). Bauxite residue issues: III. Alkalinity and associated chemistry. *Hydrometallurgy*, *108*(1–2), 60–79.

International Aluminium Institute (IAI). (2006). Aluminum technology roadmap. Retrieved May 19, 2011, from http://www.world-aluminium.org/UserFiles/File/AluminaTechnologyRoadmap%20Update%20FINAL%20May%202006.pdf

International Aluminium Institute (IAI). (2009). Aluminium for future generations/2009 update. Retrieved May 19, 2011, from http://www.world-aluminium.org/cache/fl0000336.pdf

International Aluminium Institute (IAI). (2011). Aluminium for future generations: Sustainability. Retrieved May 19, 2011, from http://www.world-aluminium.org/?pg=49

Klauber, Craig; Gräfe, Markus; & Power, Greg. (2011). Bauxite residue issues: II. Options for residue utilization. *Hydrometallurgy*, *108*(1–2), 11–32.

Margolis, Nancy. (1997). *Energy and environmental profile of the US aluminum industry*. Columbia, MD: Energetics Inc.

Martchek, Kenneth J. (2006). Modelling more sustainable aluminium. *International Journal of Life Cycle Assessment*, *11*(1), 34–37.

Plunkert, Patricia A. (2006). Aluminum recycling in the United States in 2000 (US Geological Survey Circular 1196-W). Retrieved June 28, 2011, from http://pubs.usgs.gov/circ/c1196w/

Power, Greg; Gräfe, Markus; & Klauber, Craig. (2011). Bauxite residue issues: I. Current management, disposal and storage practices. *Hydrometallurgy*, *108*(1–2), 33–45.

Roskill Information Services Ltd. (2008). *The economics of bauxite & alumina* (7th ed.). London: Roskill Information Services Ltd.

Szandelszky, Bela; & Gorondi, Pablo. (2010, October 5). Hungary sludge flood called "ecological disaster." Associated Press.

Turton, Hal. (2002). *The aluminium smelting industry*. Canberra: The Australia Institute.

Wallace, Donald H. (1937). *Market control in the aluminum industry*. Cambridge, MA: Harvard University Press.

Cap-and-Trade Legislation

One approach to controlling greenhouse gas emissions, "cap-and-trade" places caps, or limits, on allowable emissions (set by the government) but permits participating facilities to buy and sell, or trade, "allowances" on the open market. The most prominent example in US law is the acid rain control program in the 1990 Clean Air Act Amendments. Cap-and-trade legislation is now being promoted to help mitigate global climate change.

Cap-and-trade legislation is intended to directly reduce emissions of a particular pollutant into the environment. The trading feature of this legislation is attractive to economists and others because it can significantly reduce costs. Cap-and-trade programs are a prominent feature of existing and proposed climate change mitigation efforts.

Acid Rain Control

The most prominent example of cap-and-trade legislation in the United States is the acid rain control subtitle in the Clean Air Act Amendments of 1990. This legislation, which reduced sulfur dioxide emissions at a fraction of the projected cost, has been used to support the use of cap-and-trade provisions in international and national climate change programs.

The Clean Air Act was adopted originally in 1970 and has resulted in significant reductions in a variety of pollutants, including particulates, nitrogen oxides, and lead. The primary focus of the Clean Air Act is to ensure that the level of these pollutants in the ambient or outdoor air does not exceed a level that is "requisite to protect public health." Once pollutants fall to the ground or onto the water, they are not of concern under this part of the Clean Air Act. The problem, of course, is that sulfur dioxide and other pollutants settling onto the ground or surface waters acidify these environments, impair fish and insect life, and damage soil, forest, and crop productivity. The cap-and-trade provisions in the subsequent 1990 amendments were intended to reduce the total amount of sulfur dioxide that is emitted or loaded into the environment (not just the atmospheric concentration of that pollutant). And they were expected to do that as cheaply as possible.

The 1990 amendments required coal-fired electric-generating plants in the Midwest and Northeast to reduce their sulfur dioxide emissions by roughly 50 percent between 1990 and 2000. (A coal-fired power plant burns coal to generate electricity.) The legislation did that by setting a cap on overall emissions for 2000 that was roughly half the 1990 level. This cap was translated into corresponding caps for individual facilities.

Electric-generating facilities were not required by law to meet their cap—their mandatory lower level of emissions—in any particular way. They could install conventional air pollution controls, "wash" coal (removing impurities) before burning to reduce sulfur dioxide emissions, switch to less sulfur-intensive fuels (for example, from coal to natural gas), install a more efficient boiler, encourage energy savings by their customers, or do something else. The choice was entirely up to the operators of these facilities, and the US Congress expected that they would choose the cheapest means to do so.

In addition to these options, they could trade, or buy and sell, allowances. (An allowance is permission to emit, for instance, one ton of sulfur dioxide in a calendar year.) A basic premise of the legislation is that the costs of sulfur dioxide control—measured in dollars per ton of avoided emissions—vary from plant to plant. Thus some facilities may be able to reduce their emissions for fewer dollars per ton than other plants, and they may be able to do so even while reducing their emissions below the required cap. Under the 1990 amendments, the plant with the lower control costs can sell its excess emissions—in the form of allowances—to the plant with higher control costs.

Consider two plants, A and B, both of which have a cap that is 100 tons below current levels. Control costs at plant A are $30 per ton, and control costs at plant B are $70 per ton. Without trading, costs to meet the cap would be $3,000 for plant A and $7,000 for plant B, for a total of $10,000. With trading, plant A could reduce its emissions by 200 tons, and plant B could buy the "excess" reduction of 100 tons (in the form of 100 allowances) for a lot less than $7,000. (In a well-functioning market, competitive pressure will drive down the cost of allowances.) Thus plant A offsets some of its costs by selling allowances, and plant B is able to comply for a lower cost than it would otherwise pay. While cap-and-trade programs are often called trading programs, there would be no need or incentive to trade without the cap.

The 1990 amendments have worked as intended. Costs have been less than half of what was expected, and emissions were reduced more than required. During this same period, US gross domestic product and electricity generation also increased. According to a 2009 article by the Environmental Defense Fund, the program "has demonstrated that environmental protections need not compete with economic well-being."

Kyoto Protocol

In 1992, the United States became the fourth country in the world to ratify the United Nations Framework Convention on Climate Change (UNFCCC), which establishes an international legal structure to address climate change but does not establish any binding or numerical emission limits. When the parties began negotiating a protocol (a separate agreement under the convention) to set binding emission limits, the United States used its experience under the Clean Air Act to advocate trading mechanisms as a way of reducing costs. When this protocol was finalized in 1997 in Kyoto, Japan, it required developed countries to reduce their greenhouse gas emissions by 5 percent from 1990 levels by 2012. The required level of emissions reductions varied by country; the United States would have been obliged to reduce its emissions by 7 percent from 1990 levels, and western European countries (including the European Union) were required to reduce their emissions by 8 percent.

The protocol is in effect; it has been ratified by all developed countries except (ironically) the United States. (In 2001, President George W. Bush expressly repudiated the Kyoto Protocol "because it exempts 80 percent of the world, including major population centers such as China and India, from compliance, and would cause serious harm to the US economy.")

The Kyoto Protocol also contains several different emissions trading provisions to decrease the costs of those reductions. The Kyoto Protocol allows emissions trading between developed countries. As a result, these parties can take advantage of cost differences that exist between them (Article 17). The protocol also allows another form of emissions trading between developed countries, known as "joint implementation." Under joint implementation (Article 6), a developed country enters an agreement with another developed country to carry out an emissions reduction project in that other country and counts those reductions toward its own obligations. The protocol also created the clean development mechanism (CDM), an innovative kind of trading program (Article 12). The CDM allows developed countries to receive credit toward their required reductions for any reductions based on projects undertaken in *developing* countries. The twin purposes of the CDM are to assist developing countries "in achieving sustainable development" and to help developed countries meet their emissions reduction requirements. The CDM is particularly attractive because of the great difference between costs in developed and developing countries. One consequence of the Kyoto Protocol is the European Union's development and implementation of an emissions trading system.

The use of trading mechanisms in an international setting has raised a number of implementation and methodology issues:

- For both joint implementation and the clean development mechanism, the emissions reductions from the project must be "additional to any that would occur in the absence of the certified project activity" (Article 12.5(c)). It has often been hard in practice to determine that. As a consequence, credit may be given for reductions that would have occurred anyway.
- For many projects, such as carbon dioxide reductions that are achieved through forestry, it is difficult to directly determine the exact reductions that are being achieved. Reductions are calculated based on models and projections, which may or may not be reasonably accurate. By contrast, the power plants subject to trading in the United States are subject to continuous emissions monitoring, so it is relatively easy to determine the actual reductions being achieved.
- The US legal system contains a great many mechanisms assuring the integrity and enforceability of legal agreements, including the regulatory system on which they are based. In many developing countries, legal systems are less well developed and corruption is widespread. In consequence, it may be harder to enforce agreements

for reductions or guarantee the integrity of reductions claimed to be achieved in those countries.

At the end of 2009, international negotiations were being conducted for a successor agreement (or agreements) to the Kyoto Protocol. A successor agreement is needed because the compliance period for the Kyoto Protocol ends in 2012, and there is no international agreement for emissions reduction after that. One possibility is a second cap-and-trade agreement built on Kyoto that requires further reductions in greenhouse gas emissions.

European Union Emissions Trading System

The European Union has established the world's first international trading system for carbon dioxide emissions—the European Union Emissions Trading System (EU ETS). The system was established to help EU Member States comply with their Kyoto Protocol commitments. The EU ETS is set up to operate in three phases, or "trading periods." The first period, a trial period, ran from 2005 to 2007. Its purpose was not to meet Kyoto targets but to gain experience with emissions trading. The second period continues from 2008 to 2012, which coincides with the Kyoto Protocol. The third period will extend from 2013 to 2020.

Within the EU, the system raises several important implementation questions. One is whether the EU or member states will allocate allowances. Under the EU ETS, member states allocate allowances in the first and second periods, while the EU will do so in the third period. Another issue is whether allowances are allocated for free or auctioned. While the bulk of allowances are now allocated for free, there is a movement toward greater use of auctioning. Still another question is whether enforcement is done by the EU or member states. Although the system is EU-wide, enforcement is undertaken by member states. According to the European Environmental Agency (EEA), EU emissions in 2008 declined for the fourth consecutive year. If planned and existing measures are fully implemented, and member states take advantage of the clean development mechanism and other provisions of the Kyoto Protocol, the EEA projects that the EU will overachieve the 8 percent reduction target set in the Kyoto Protocol.

US Regional Initiatives

Ten northeastern and mid-Atlantic states participate in the Regional Greenhouse Gas Initiative (RGGI), which has developed a model rule to establish a cap-and-trade program for electric utilities. The ten states are: Connecticut, Delaware, Maine, New Hampshire, New Jersey, New York, Vermont, Massachusetts, Rhode Island, and Maryland.

The overall environmental goal for RGGI is for each state to adopt a carbon dioxide trading program for emissions from fossil fuel–fired electricity-generating units having a rated capacity equal to or greater than 25 megawatts. These states together have negotiated a model rule that is being used, in each state, as the basis for the program. Power plants are an attractive starting point because they have already experienced the sulfur dioxide trading program under the Clean Air Act. Emissions reductions are to occur from 2015 to 2018, at a rate of 2.5 percent annually for each of the four years. By 2018, each state's base annual emissions budget is to be 10 percent below its initial budget.

Similarly, the Western Climate Initiative (WCI) involves a regional emissions cap for multiple economic sectors and a cap-and-trade system. The WCI is comprised of seven western States (Arizona, California, New Mexico, Montana, Oregon, Utah, and Washington) and four Canadian provinces (British Columbia, Manitoba, Ontario, and Quebec). The goal of the WCI is to reduce greenhouse gas emissions by 15 percent from 2005 levels by 2020.

Proposed US Legislation

As of January 2010, comprehensive climate change legislation, based on cap-and-trade, is nearing passage in Congress. On 26 June 2009, the House of Representatives passed the American Clean Energy and Security Act (H.R. 2454). On 5 November 2009, the Senate Environment and Public Works Committee approved a somewhat similar bill, Clean Energy Jobs and American Power Act (S. 1733).

The heart of both bills is a cap-and-trade program for greenhouse gas emissions. The House bill would require the United States to reduce its greenhouse gas emissions by 83 percent from 2005 levels by 2050 (which equates to a 69 percent cut from 1990 emissions levels by 2050).

Both bills would create a cap-and-trade program for "covered entities." The term includes all electric power plants as well as factories and other facilities that produce 25,000 tons of carbon dioxide or carbon dioxide equivalent (gases such as methane and nitrous oxide). These facilities are responsible for about 85 percent of US greenhouse gas emissions.

These reduction requirements create an emissions cap or limit for covered facilities, and the level of this cap declines over time. Covered facilities can meet this cap more or less as they see fit—for example, by becoming more energy efficient, switching to a less carbon-intensive fuel (from coal to natural gas), or using more renewable energy. Another option for covered facilities is trading, or purchasing emissions allowances. Like sulfur dioxide controls under the Clean Air Act, some facilities will be able to reduce their greenhouse gas emissions on a per-ton basis more cheaply than others. Those that do can trade or sell their "excess" reductions—in the form of allowances that are equal to one ton of carbon dioxide or carbon dioxide equivalent—to facilities where control costs are greater.

A cap-and-trade system such as that contained in these bills should lead to a price on carbon that would have ripple effects throughout an economy. (Though politically less likely, a carbon tax would have the same effect.) The price would be reflected in the market price for allowances. According to conventional economic wisdom, the economic pressure created by a cap-and-trade program should lead to less use of fossil fuels, greater use of less-carbon intensive fuels, and other changes that would result in lower greenhouse gas emissions.

The House and Senate bills contain a great many other provisions, to be sure. Many of these are directed at making sure that the emissions trading market is transparent, reliable, and functions smoothly. Others are directed at ensuring that the price of allowances doesn't get so high that the program becomes unaffordable for many facilities. Some provisions allow covered facilities to purchase "offset allowances" in the United States, primarily from foresters and farmers, to meet their emissions caps. Offset allowances are allowances generated by non-covered facilities in the form of reduced greenhouse gas emissions or increased carbon sequestration or storage. The bills would also establish a system of national emissions reporting.

For climate change legislation, cap-and-trade provisions raise the following issues:

- How to allocate allowances. Under the Clean Air Act, allowances are allocated for free. Under RGGI, many states are auctioning some or all of the allowances. Utilities prefer free distribution of greenhouse gas allowances because, they say, it reduces their costs. It may also make it easier to pass the legislation. The argument for auctioning, by contrast, is based on the idea that the government shouldn't give away something of value, and that it should not give an economic advantage to existing companies.
- Distribution of proceeds from the sale of allowances. To the extent allowances are sold rather than given away, the government could use or distribute the money for a variety of purposes—including fostering energy efficiency, retraining workers who are adversely affected, and research and development. Alternatively, some or all of the proceeds could be distributed to individual taxpayers as a kind of rebate, either on a one-time or continuing basis.
- How to treat existing regional initiatives such as RGGI and WCI. While there is understandable interest in using federal legislation to create a national cap-and-trade system, many believe that federal legislation should accommodate the time and investments already made in those programs.
- The extent to which cap-and-trade measures need to be supplemented with other rules. Because of market imperfections, the economic pressure caused by a cap-and-trade program will not always have the desired result. According to the economist Robert Stavins (2007), consumers often do not purchase products that are more energy efficient because they undervalue the economic savings of those products. In addition, the person with the ability to achieve greater energy efficiency (for example, the landlord) is frequently not the person who pays the energy bills (typically the tenant). The incentive, in other words, writes Stavins, is not directed at the person with the ability to make decisions that will reduce greenhouse gas emissions. Beyond that, he says, the price signal provided by a cap-and-trade or tax program is not likely to lead to sufficient investment in the variety of different research and development activities needed to mitigate climate change. Finally, while a cap-and-trade program can surely reduce the costs of emissions control, it is less likely to lead to more immediate environmental, social, and economic co-benefits than a performance standard of equivalent stringency. According to David Driesen, an author and professor of law, experience under the Kyoto Protocol shows that buyers of emissions allowances are primarily interested in reducing their costs, not in fostering or capturing the other benefits that may come from a use of a particular policy or measure (Driesen 2008). These limitations in a stand-alone cap-and-trade program strengthen the case for energy-efficiency policies, for measures that would drive greater levels of private investment, and for programs that would generate substantial economic, social, and environmental benefits in addition to greenhouse gas reductions.

These issues are addressed, to some degree, in two manners. First, other federal laws already indirectly address greenhouse gas emissions in ways that would complement federal cap-and-trade legislation. The Energy Independence and Security Act of 2007, for example, requires automobiles and light trucks (including sport-utility vehicles) to achieve a combined standard for model year 2020 of at least thirty-five miles per gallon. Second, while the comprehensive climate change bills before Congress include cap-and-trade provisions, they also include a great many other provisions, some of which address these issues. The House bill, for instance, would establish a national energy efficiency program for buildings, would strengthen existing requirements for energy efficient lighting and appliances, and would foster energy efficiency in transportation and at industrial facilities. Still it is not clear whether such provisions would overcome the limitations of cap and trade.

Beyond these issues, there is a lingering question about whether a carbon tax would be a better approach. Cap-and-trade legislation would achieve a definite reduction because of the cap. By contrast, the emissions reduction effect of a tax is difficult to determine in advance. On the other hand, a tax is more economically efficient because it would apply to all sources of greenhouse gas emissions, rather than just those identified in the legislation itself. The comparative cost of the two programs would depend on their design details—how allowances are allocated and similar issues. A 2008 survey of economists by the General Accountability Office found that eleven favored cap and trade while seven favored a tax.

Implications

Cap-and-trade legislation has made it possible to achieve greater reductions of many pollutants, for a lower cost, than was previously thought possible. While it is understood to be a market mechanism for improving environmental quality, it is important to recognize that an old-fashioned "command-and-control" rule—capping emissions from each covered facility—provides the impetus for trading. Cap-and-trade legislation is thus best understood as a blend of traditional regulation with market-based regulation. It is an effective and indispensable tool in the quest for sustainable development.

John C. DERNBACH
Widener University Law School

See also in the *Berkshire Encyclopedia of Sustainability* Automobile Industry; Airline Industry; Climate Change Disclosure; Development, Sustainable; Energy Efficiency; Energy Industries (*assorted articles*); True Cost Economics

Further Reading

Dernbach, John C., & Kakade, Seema. (2008). Climate change law: An introduction. *Energy Law Journal, 29*(1), 12–14.

Driesen, David M. (2008). Sustainable development and market liberalism's shotgun wedding: Emissions trading under the Kyoto Protocol. *Indiana Law Journal, 83*(21), 52–57.

Environmental Defense Fund. (2008). The cap and trade success story. Retrieved July 5, 2009, from http://www.edf.org/page.cfm?tagID=1085

Environmental Protection Agency. (1990). Clean Air Act: Title IV—acid deposition control. Retrieved July 5, 2009, from http://www.epa.gov/air/caa/title4.html

European Environment Agency. (2009). *Greenhouse gas emission trends and projections in Europe 2009: Tracking progress towards Kyoto targets.* Retrieved December 31, 2009, from http://www.eea.europa.eu/publications/eea_report_2009_9

Faure, Michael, & Peeters, Marjan (Eds.). (2008). *Climate change and European emissions trading: Lessons for theory and practice.* Northampton, MA: Edward Elgar.

Government Accountability Office. (2008, May). *Climate change: Expert opinion on the economics of policy options to address climate change.* Retrieved July 5, 2009, from http://www.gao.gov/new.items/d08605.pdf

Stavins, Robert N. (2007). *Proposal for a US cap-and-trade system to address global climate change: A sensible and practical approach to reduce greenhouse gas emissions.* Retrieved July 5, 2009, from http://ksghome.harvard.edu/~rstavins/Papers/Stavins_Hamilton_Working_Paper_on_Cap-and-Trade.pdf

United Nations Framework Convention on Climate Change (UNFCCC). (1997). Essential background. Retrieved July 16, 2009, from http://unfccc.int/essential_background/items/2877.php

US Senate Committee on Environment & Public Works. (2008). Lieberman-Warner Climate Security Act of 2008. Retrieved July 5, 2009, from http://epw.senate.gov/public/index.cfm?FuseAction=Files.View&FileStore_id=aaf57ba9-ee98–4204-882a-1de307ecdb4d.

Yacobucci, Brent D.; Ramseur, Jonathan L.; & Parker, Larry. (2009). *Climate change: Comparison of the cap-and-trade provisions in H.R. 2454 and S. 1733.* Retrieved December 31, 2009, from http://assets.opencrs.com/rpts/R40896_20091105.pdf

Carbon Footprint

Carbon footprints represent the amount of greenhouse gases released into the atmosphere as a result of day-to-day activities such as driving, generating electricity, or manufacturing products. While footprinting methods continue to evolve, the indicator has value especially for monitoring activity and encouraging more sustainable behavior at all levels, from the individual to the global society.

A carbon footprint is an estimation of an entity's contribution of greenhouse gases (GHGs) to the global atmosphere. Carbon footprints can be estimated for individuals, cities, countries, products or processes, organizations, and ultimately, for the global population. The release of GHGs into the Earth's atmosphere is a primary driver of global climate change. Policy makers need a way to estimate human-caused GHG releases in order to assess progress toward climate stabilization goals, such as those stated in the Kyoto Protocol of 1997, and the concept of the carbon footprint serves that purpose.

Carbon footprint is used as shorthand for *greenhouse gas footprint* because many GHGs are carbon based, including carbon dioxide (CO_2), methane (CH_4), hydrofluorocarbons (HFCs), and perfluorocarbons (PFCs). For comparison and analysis, GHG releases are converted into equivalent quantities of carbon dioxide (CO_2e) using an estimate for each gas of its potential to cause global warming.

The footprint aspect of the term is derived from the ecological footprint, which measures the natural resources needed to sustain human activities. If our ecological footprint is larger than the stock of available resources, then our behavior is not sustainable over time. A similar premise applies to the carbon footprint; if GHG releases are more than the environment can absorb without fundamentally altering the global climate system, then our behavior is unsustainable. A technical challenge lies in knowing what level of releases the environment can absorb without triggering catastrophic climate change, and at what level carbon footprints become unsustainable. For this reason, we tend to think simply: smaller is better.

Measurement Considerations

GHG releases are estimated from activities rather than directly observed by pollution monitors. Standardized protocols have been developed to estimate GHG releases for countries, leading to relatively comparable data by country. These protocols include the six major gases specified in the Kyoto Protocol, a standard procedure for converting GHGs to CO_2e, and a list of economic sectors and source activities that should be included in the assessments. The protocols are, however, based on estimates of activities and estimates of the releases from those activities, which are oversimplified reflections of reality.

To estimate the carbon footprint from driving a motor vehicle, for instance, one would estimate the amount of vehicle miles traveled (by vehicle type, such as passenger car, small truck, and large truck), estimate the amount of GHG released when driving (by vehicle type), and then multiply these amounts together (and sum across vehicle types) to get an estimate of GHG releases from driving. Both the estimation of vehicle travel and the estimation of emissions from vehicles are imprecise. Such imprecision compounds as footprints are calculated for multiple activities, including other types of energy fuel combustion, industrial processes and product use, agriculture, and waste disposal. As scientific knowledge improves, protocols must be updated and prior measurements recalculated, although the accuracy of estimates is always limited by data availability.

One consideration involves emissions from land-use change, such as the conversion of forest into agricultural or urban land. GHGs can be absorbed by the oceans and by vegetation on land. What is not absorbed by oceans or vegetation builds up in the atmosphere. A country's footprint can be adjusted downward to account for the extent to which available vegetation can absorb GHG releases, or adjusted upward if activities such as deforestation reduce the capacity of vegetation to absorb GHGs.

Many footprints entirely exclude emissions from land-use change, or report the footprints with and without land-use change. Data regarding emissions from land-use change tend to be available only for large geographies such as continents, be based on infrequent data collection, and contain substantial uncertainties when apportioned to smaller geographies. Available data suggest that tropical countries with extensive deforestation like Brazil and Indonesia release the majority of GHGs from land-use change. Alternatively, urbanizing countries like China and the United States have seen their vegetative capacity increase in recent years as agriculture is abandoned and the landscape transitions to denser vegetation. The inclusion of land-use change emissions opens the door to reduction strategies focused on reforestation.

Another consideration involves indirect emissions. The international protocol focuses on measuring GHG emissions where they are released, using "production-based" accounting. This approach is the easiest to accomplish given available activity data. An issue arises when releases within the country boundaries are related to production of goods or services consumed elsewhere. Much global industrial activity has moved to developing nations, even while most consumers of those industrial goods remain in developed nations. Similarly, disposal of waste has shifted to developing nations. Indirect GHG emissions associated with product consumption and waste disposal in developed nations may be substantially larger than is apparent from direct GHG releases within their borders.

For this reason, many scholars advocate using "consumption-based" accounting, in which the production-based footprint is (roughly) adjusted upward in high-consumption areas and adjusted downward in high-production areas. Calculation of consumption-based footprints requires analysis of environmental input and output or analysis of the life cycle of a product, tracking production, consumption, and disposal activities across spatial boundaries and over time. Data requirements are extensive. Advances have been made in calculating footprints of organizations and products using a life cycle approach, driven in large part by a desire to verify sustainable business practices. The methods are still being refined for countries and smaller geographies, especially to avoid double counting of GHG emissions at the point of production and the point of consumption, such as for electricity. Interim methods estimate a partial footprint associated with consumption from such activities as driving and heating homes.

The choice of production-based or consumption-based accounting has implications for policy. Assigning GHGs to the locations where they are released implicitly places responsibility for GHG reduction on producers, following the principle that the polluter pays, while ignoring consumer demand. This places the focus on obtaining reductions from the largest global contributors, such as China, the United States, Europe, Russia, and India. Given the spatial distribution of production activities, responsibility could fall on countries unable to afford reductions without compromising economic development, such as China and India. Policy makers question whether affluent areas like the United States and the European Union should shoulder more of the GHG reduction burden given their historical extensive releases and consumption activities, following a principle of shared responsibility.

A similar issue arises when calculating carbon footprints for urban areas, which have diverse combinations of production and consumption activities. Estimation of city footprints is complicated by the lack of a standard definition of *urban* and the lack of subnational data on activity and emission rates. As of 2011, most efforts focus on megacities, where data are more available, or on cities within single regions. Differences in methodologies make footprints difficult to compare across cities and over time.

Progress has been made at estimating consumption-based footprints for select global cities. A protocol for calculating carbon footprints of public sector organizations within the United States was released in 2010, following a similar approach (WRI and LMI 2010). According to the Greenhouse Gas Protocol, organizations are expected to report emissions from activities within their operational control and emissions from activities produced elsewhere but consumed within the organization's boundaries, such as purchased electricity. The organizations have discretion over including other emissions, such as for waste disposal and cross-boundary transportation, which can improve the comprehensiveness of the footprint but limit comparability across entities.

Estimates and Implications

The most widely available information on carbon footprints at the country level is production based. Footprints may be represented in three ways. (See table 1 below.) The broadest representation is in terms of a country's total GHG releases in a given year. A second figure is a country's footprint in terms of GHG releases per person per year, as a way to adjust for the size of a country; countries with larger populations are likely to have larger footprints. A third measurement is a country's footprint in terms of GHG releases per unit of economic output per year (known as emissions intensity), as a way to adjust for a country's development status; more developed countries are likely to have larger footprints, although their affluence may allow them to utilize low-carbon technologies and reduce their footprints.

China had the largest carbon footprint of the five areas listed in table 1, releasing nearly 20 percent of the global GHGs in 2005. The sheer size and rapid growth of emissions from China makes its footprint globally unsustainable. China's emissions were, however, spread across a population of nearly 1.3 billion residents, resulting in a much smaller footprint than, for example, the United States when viewed on a per person basis. Using the per person measure, which is most common for comparing footprints, it appears that Americans live much less sustainably than do residents of the Russian Federation, European Union, China, or India. Alternatively, when using the emissions intensity indicator, it appears from the data in table 1 that the United States' footprint is more sustainable than that of India, China, or the Russian Federation but less sustainable than that of the European Union. Although the hope is to see all three footprint measures decline over time, total GHG releases matter most with regard to climate sustainability.

The three different approaches for representing footprints remind us that carbon footprints are determined by complex interactions of population, affluence, and available technology. Globally, rapid population growth and continued pressure to improve living standards are expected across much of the developing world, which will place upward pressure on footprints. Achieving climate stabilization goals thus requires focusing on reducing emissions intensity. Transitions to low- and no-carbon energy technology, including renewable fuels and nuclear power, could allow countries to develop economically while reducing their carbon footprint. Such transitions would be expensive and are a source of contention internationally, as nations such as China seek to utilize cheap, available, and carbon-intensive fuels like coal. In addition, technology improvements alone may simply increase the total amount of fuels consumed and carbon released unless they are coupled with other restraints on consumption. Reductions in consumption may be particularly important in high-income countries and cities, where affluence drives consumption and consumption dominates production.

Future Directions

The methods used to estimate carbon footprints are constrained by available data and rely on assumptions and aggregations that may not reflect reality well. Advances are needed in estimating releases for GHGs other than carbon dioxide, in estimating activity at smaller geographic scales, and in making adjustments to footprints

TABLE 1. Greenhouse Gas Releases for Selected Areas in 2005

	Total GHG Releases (million metric tons CO_2e)	**GHG Releases Per Person (metric tons CO_2e)**	**Emissions Intensity (metric tons CO_2e per $GDP)**
China	7,232.8	5.5	1,361.0
United States of America	6,914.2	23.4	559.2
European Union	5,043.1	10.3	382.5
Russian Federation	1,954.6	13.7	1,151.2
India	1,859.0	1.7	760.3

Notes: Higher values are less sustainable; emissions from land-use change and forestry are excluded.

Source: World Resources Institute (2011).

to reflect consumption behavior rather than production behavior. Differences in methodology make footprints difficult to compare across space and time.

Even without advances, the footprint concept has utility for policy makers for monitoring changes in behavior over time and examining the effectiveness of GHG reduction strategies in particular locations. The footprint concept also has financial utility for estimating the GHGs not released when reduction strategies are pursued, which can be bundled into credits and traded on a market. Measurement of one's carbon footprint by an individual or business also spreads awareness and signals a commitment to sustainable behavior, which can be a powerful force for curtailing GHG releases. For these reasons and more, the carbon footprint is likely to remain a popular indicator of climate sustainability for countries, cities, businesses and organizations, products and processes, and individuals.

Andrea SARZYNSKI
University of Delaware

See also in the *Berkshire Encyclopedia of Sustainability* Air Pollution Indicators and Monitoring; Computer Modeling; Ecological Footprint Accounting; Energy Labeling; Human Appropriation of Net Primary Production (HANPP); I = P × A × T Equation; Reducing Emissions from Deforestation and Forest Degradation (REDD); Regulatory Compliance; Shipping and Freight Indicators; Tree Rings as Environmental Indicators

FURTHER READING

Bader, Nikolas, & Bleischwitz, Raimund. (2009). Measuring urban greenhouse gas emissions: The challenge of comparability. *Surveys and Perspectives Integrating Environment and Society, 2*(3), 1–15.

Blake, Alcott. (2005). Jevons' paradox. *Ecological Economics, 54*(1), 9–21.

Brown, Marilyn A.; Southworth, Frank; & Sarzynski, Andrea. (2008). *Shrinking the carbon footprint of metropolitan America*. Washington, DC: The Brookings Institution.

Carbon Trust. (2010). Carbon footprinting: The next step to reducing your emissions. Retrieved November 14, 2011, from http://www.carbontrust.co.uk/Publications/pages/publicationdetail.aspx?id=CTV043

Dodman, David. (2009). Blaming cities for climate change? An analysis of urban greenhouse gas emissions inventories. *Environment & Urbanization, 21*(1), 185–201.

Hoffert, Martin I., et al. (1998). Energy implications of future stabilization of atmospheric CO_2 content. *Nature, 395*, 881–884.

Houghton, Richard. (n.d.). Data note: Emissions (and sinks) of carbon from land-use change. Retrieved November 14, 2011, from http://cait.wri.org/downloads/DN-LUCF.pdf

Intergovernmental Panel on Climate Change (IPCC). (2006). 2006 IPCC guidelines for national greenhouse gas inventories. Hayama, Japan: IPCC.

Kaya, Yoichi. (1989). *Impact of carbon dioxide emission control on GNP growth: Interpretation of proposed scenarios*. Hayama, Japan: IPCC.

Kennedy, Christopher A.; Ramaswami, Anu; Carney, Sebastian; & Dhakal, Shobhakar. (2009, June 28–30). Greenhouse gas emission baselines for global cities and metropolitan regions (paper, Proceedings of the 5th Urban Research Symposium). Marseilles, France.

Lenzen, Manfred; Murray, Joy; Sack, Fabian; & Wiedmann, Thomas. (2007). Shared producer and consumer responsibility: Theory and practice. *Ecological Economics, 61*, 27–42.

Levinson, Arik. (2010). Offshoring pollution: Is the United States increasingly importing polluting goods? *Review of Environmental Economics and Policy, 4*(1), 63–83.

Marcotullio, Peter J.; Sarzynski, Andrea; Albrecht, Jochen; Schulz, Niels; & Garcia, Jake. (2012). Assessing urban GHG emissions in European medium and large cities: Methodological issues. In Pierre Laconte (Ed.), *Assessing sustainable urban environments in Europe: Criteria and practices*. Lyon, France: Centre for the Study of Urban Planning, Transport and Public Facilities.

Satterthwaite, David. (2008). Cities' contribution to global warming: Notes on the allocation of greenhouse gas emissions. *Environment & Urbanization, 20*(2), 539–549.

Wackernagel, Mathis, & Rees, William F. (1996). *Our ecological footprint: Reducing human impact on the Earth*. Gabriola Island, Canada: New Society Publishers.

Wiedmann, Thomas, & Minx, Jan. (2008). A definition of "carbon footprint." In Carolyn C. Pertsova (Ed.), *Ecological economics research trends*. Hauppauge, NY: Nova Science Publishers.

World Resources Institute (WRI). (2011). Climate Analysis Indicators Tool (CAIT) version 8.0. Retrieved November 14, 2011, from http://cait.wri.org/

World Resources Institute (WRI) & Logistics Management Institute (LMI). (2010). The Greenhouse Gas Protocol for the US public sector. Retrieved November 28, 2011, from http://www.ghgprotocol.org/files/ghgp/us-public-sector-protocol_final_oct13.pdf

Coal

Coal continues to be an important energy and industrial resource. To make the coal industry sustainable, new technologies must address the emissions and poor coal utilization that occur during mining, transport, and conversion into power and chemicals. Potential sustainable technologies should: improve coal conversion; reuse by-products and waste; develop efficient combustion, power generation, and gasification technologies; and implement carbon capture, storage, and utilization strategies.

Coal is widely used in the power, iron and steel, chemical, and construction materials industries, and in human daily life. Historically, its use was advantageous as it prevented forests from being cut down for firewood and charcoal. But the coal industry consumes resources and affects the environment in numerous negative ways. Coal is a natural resource that is nonrenewable and nonrecyclable, and its consumption for fuel leads to higher emissions of acidic rain gases, heavy metals, particulates, and carbon dioxide into the Earth's atmosphere.

Coal, accounting for about 30 percent of the world's total primary commercial energy supply, is likely to remain a key energy resource into the mid twenty-first century. The main challenges to the coal industry pertain to utilization efficiency and the emissions of various pollutants during mining, transport, and consumption as fuel. Although many advanced technologies have been developed and deployed to address these problems, coal utilization industries still suffer from low efficiency and high pollution.

Global Resources

Coal exists as anthracite, which is nearly pure carbon and is commonly used for commercial and residential heating; bituminous coal, which is less pure and is used primarily for coke production and to fuel steam-electric power generation; and brown coal or lignite, which is highest in compounds that volatilize on burning, is used for power generation plants, and makes up the majority of remaining coal reserves. The United States has the largest coal reserves in the world (27.6 percent of global reserves), with Russia (18.2 percent) and China (13.3 percent) following (BP 2011). China, however, has dominated coal production since 1990, when it overtook the United States to become the world's largest coal producer. In 2010 China produced 48.3 percent of coal in the world. This was more than three times the amount of coal produced in the United States, the second highest producer. Both China and the United States consume 95 percent of the coal they produce, which together is 63 percent of global coal consumption (BP 2011).

Use and Processing

Coal is a "dirty" solid energy and carbon resource and is more difficult to use than natural gas and petroleum oil. It contains ash and many pollutant elements such as sulfur, nitrogen, halogen, and heavy metals (e.g., mercury, arsenic, and chromium). Compared with biomass fuels made from plants, coal is not renewable and is not carbon neutral because nothing offsets the carbon it releases when combusted.

Coal is primarily used as an energy resource to produce heat and electricity through combustion. In China more than 50 percent of the coal consumed annually is burned to generate power (Market Avenue 2008). Coal generally is pulverized and then combusted in a furnace with a boiler, where the furnace heat converts the boiler water to steam, which is then used to spin steam turbines that turn generators to produce electricity. Although the thermodynamic efficiency of this process has improved, conventional steam turbines with the most advanced technology have topped out at 37–38 percent efficiency for the entire process (combustion plus power generation). A supercritical turbine, which runs a boiler at an extremely high temperature and pressure—so high that the water in the boiler remains in liquid state—can realize a heat efficiency of 42 percent, while an ultrasupercritical turbine (with an even higher temperature and pressure) can achieve a heat efficiency of 45 percent or greater (World Coal Institute 2009).

Coal can be used as a raw material to produce coke, a solid, carbon-rich residue left after the volatile components (those that burn or evaporate at low temperatures) are removed from bituminous coal. Coke has many industrial uses; for example, in metallurgical processes, it is used as either a fuel or a reducing agent to provide heat and to smelt iron ore in a blast furnace.

Coal can also be used as a feedstock, or a carbon resource, in chemical production. Coal gasification converts coal to gaseous products composed of hydrogen, carbon dioxide, carbon monoxide, methane, and other hydrocarbons. Coal can be gasified to produce syngas, a mixture of carbon monoxide and hydrogen, which can be converted synthetically to chemicals such as methanol, dimethyl ether (DME), or ammonia, or to transportation fuels such as gasoline or diesel. Alternatively the hydrogen obtained from gasification can be used for various purposes such as operating combined cycle power-generating plants, driving fuel cells, and upgrading fossil fuels.

Because coal will continue to be an important resource, cleaner coal technologies are required to ensure energy security and increase the coal industry's sustainability. One developing approach is carbon capture and storage, also called carbon capture and sequestration (CCS). This strategy proposes to significantly reduce the carbon dioxide that is released by burning coal, petroleum, and natural gas and thus stabilize greenhouse gas concentrations. CCS involves trapping the excess mixture of carbon dioxide and toxic gases formed when burning or gasifying coal and separating out the carbon dioxide. The gas is then placed in long-term storage by, for example, injection into depleted petroleum production fields, deep saline aquifers (geological reservoirs containing brackish or salt water), deep coal seams geologically too thin to mine, or the oceans (with consideration to environmental safety).

Other advanced technologies are needed as well for meeting the rapidly increasing energy demand and for cost-effectively reducing greenhouse gas emissions (primarily carbon dioxide). Technological opportunities for supporting the industry's sustainable development exist in coal mining, transportation, conversion of coal to power or chemicals, and in iron, steel, and construction materials production.

Mining

There are two primary ways to mine coal, surface (or open-cut) mining and underground (deep) mining. Surface mining extracts seams of coal that are near the surface by removing the overburden (the rock and soil that is above the coal) to access the coal from the above ground. Surface mining can destroy ecosystems and dramatically alter a region's topography. In the United States, for example, the surface mining technique known as mountaintop removal directly destroyed more than 2000 square kilometers of land between 1985 and 2001 and left a much greater area damaged (US EPA 2003). This technique transfers the rock and soil from mountain summits to adjacent valley streams, creating barren plateaus from forested mountain ridges while polluting and destroying streams.

The Coal Mining Task Force of the Asia-Pacific Partnership on Clean Development and Climate (2006, 54) estimates that about 40,000 new hectares (400 km^2) are disturbed by coal mining activities each year in China alone (this will span the area of Switzerland every ten years). Although regulations for land reclamation exist in most mining countries, they are often "piecemeal and less efficient" in developing countries (Xia 2006). Even in developed countries, the reclamation process—especially the removal of contaminants from the soil—is long and costly with the current technology; the United Nations Environment Programme estimates that between US$10 billion and US$22 billion are spent on land rehabilitation measures each year (Xia and Shu 2003).

Underground mining is not as severely damaging to the environment but can pose significant safety hazards for miners. China has by far the highest number of deaths caused by mining accidents in the world. In response to the deaths of thousands of miners per year in coal mining accidents, China's National Development and Reform Commission issued a five-year plan for the country's coal industry, starting in 2006, that placed stricter requirements on the security, efficiency, cleanliness, and environmental impact of coal exploration, mining, and delivery operations (Blueprint for coal sector 2007). The number of coal-worker deaths in China fell from 5,986 in 2005 to 2,631 in 2009, although some estimate these numbers could be much higher; for comparison, only 34 mining deaths were reported in the United States in 2009 (Associated Free Press 2010; Alford 2010).

Coal mining is dirty and dangerous. It disturbs the land's surface and the underground water, which damages or adversely affects a region's commerce and public life. It should be stressed that the notion of "clean coal" refers only to cleaner methods of burning coal: there is no "clean" way to mine coal. In addition to safety issues, the coal mining industry could also address sustainability and resource utilization by developing technologies that use gangue (a commercially worthless rock found with coal and minerals) in building and road construction materials, for example, and by reducing pollution associated with the transport of coal from mining sites to power plants (one such option would be to locate processing facilities closer to mining sites).

Converting Coal to Power

The amount of pollutant emissions from burning coal is much higher than from combusting oil and natural gas (US EIA 1999, 58), thus the majority of atmospheric pollutants and carbon dioxide emissions from fuel consumption are from coal combustion. (See table 1.)

Potential technologies for cleaner coal power generation include ultrasupercritical (USC) power generation, integrated gasification combined cycle (IGCC) systems, and polygeneration systems based on IGCC. USC power plants can reach 44–45 percent efficiency. Because of the higher efficiency, carbon dioxide emissions per unit of electricity at USC power plants are about one-fifth that of traditional power plants. IGCC systems use coal gasification, syngas purification, and the gas-steam combined cycle of power-generation turbines.

IGCC-based polygeneration refers to cleaner and more efficient production processes that integrate a variety of coal conversion and synthesis technologies. These create a variety of cleaner secondary energy sources (e.g., oil, gas) and value-added chemical products. Compared with traditional power generation and single-product synthesis technologies, the advantages of coal-based polygeneration systems are high efficiency, low cost, and low emissions. Opportunities for advanced IGCC systems exist in upgrading all the related technologies, including gasification, gas cleaning, gas turbines, and chemical synthesis.

Table 1. Pollutant Emissions from Burning Coal, Oil, and Natural Gas (in kilograms per billion kilojoules).

Pollutant	Coal	Oil	Natural Gas
Carbon dioxide (CO_2)	99,500	78,500	56,000
Carbon monoxide (CO)	99.5	15.8	19.1
Nitrogen oxide (NO)	218.7	214.4	44
Sulfur dioxide (SO_2)	1,240	537	0.48
Mercury (Hg)	0.01	0.0035	0

Source: United States Energy Information Administration 1999.

Coal combustion produces significantly more pollutant emissions than either oil or natural gas.

Future of Sustainable Industry Strategies

There is much room for improvement in more sustainable coal utilization strategies and technologies, as well as in mining and transport practices. Researchers should seek environmentally sustainable, affordable, high-efficiency conversion technologies, and should pursue integrated comprehensive solutions that raise efficiency and lower environmental damages.

Concrete actions include monitoring coal mine safety and efficiency and improving the investment environment for the construction and operation of advanced power plants, coking plants, and other coal chemical industry facilities. The main tasks for the coal industry include the following:

- accelerating coal conversion industries at coal mines to develop a new coal utilization mode;
- promoting efficient utilization of coal-bed methane and coal gangue to reduce the consumption of high-rank coal resources and to control pollution;
- boosting land reclamation at coal mines and using coal mine water to protect the mine's eco-environment;
- developing advanced combustion technologies to generate heat and electricity with a minimum of pollution;
- applying more supercritical and ultrasupercritical power generation technologies to raise the power industry's efficiency;
- developing large-scale, advanced gasification technologies for low-grade coal to extend resource availability and to create syngas or fuel gas; and
- advancing low-cost, clean technologies and carbon capture and storage.

In addition, process-intensification and scale-up technologies can maximize system efficiency and minimize pollution emissions. In the long term, a more sustainable coal industry could be achieved by promoting a polygeneration technology system integrated with renewable energy and carbon dioxide capture and sequestration.

ZHAO Ning, SONG Quanbin, and LIAN Ming
Shanghai Bi Ke Clean Energy Technology Co., Ltd.

XU Guangwen
Chinese Academy of Sciences

See also in the *Berkshire Encyclopedia of Sustainability* Carbon Capture and Sequestration; Greenhouse Gases; Mining—Nonmetals; Petroleum; Solar Energy; Water Energy; Wind Energy; Uranium

Further Reading

Alford, Roger. (2010, January 1). US mine deaths hit record low of 34 in 2009. Associated Press. Retrieved September 26, 2011, from http://www.newsday.com/business/us-mine-deaths-hit-record-low-of-34-in-2009-1.1679551

Asia-Pacific Partnership on Clean Development and Climate, Coal Mining Task Force. (2006). Action plan. Retrieved September 26, 2011, from http://www.asiapacificpartnership.org/pdf/Projects/Coal%20Mining%20Task%20Force%20Action%20Plan%20030507.pdf

Associated Free Press. (2010, January 20). China says coal mine deaths fall in 2009. Retrieved September 26, 2011, from http://www.chinamining.org/News/2010-01-21/1264035652d33587.html

Blueprint for coal sector. (2007, November 30). Retrieved September 26, 2011, from http://www.china.org.cn/english/environment/233937.htm

BP. (2011). Statistical review of world energy 2011. Retrieved September 20, 2011, from http://www.bp.com/sectionbodycopy.do?categoryId=7500&contentId=7068481

Cao, Yuchun; Wei, Xinli; Wu, Jinxin; Wang, Baodong; & Li, Yan. (2007). Development of ultra-supercritical power plant in China. In Kefa Cen, Yong Chi, and Fei Wang (Eds.), *Challenges of power engineering and environment: Proceedings of the International Conference on Power Engineering* 2007. pp. 231–236. Berlin: Springer.

Chang, Cheng-Hsin. (2005). Coal gasification. Retrieved September 26, 2011, from http://www.business.ualberta.ca/Centres/CABREE/Energy/~/media/University%20of%20Alberta/Faculties/Business/Faculty%20Site/Centres/CABREE/Documents/Energy/NaturalGas/ChengHsinChangCoalGasification.ashx

Freese, Barbara. (2003). *Coal: A human history*. Cambridge, MA: Perseus.

Huang Qili. (2007). Status and development of Chinese coal-fired power generation. Retrieved September 26, 2011, from http://www.egcfe.ewg.apec.org/publications/proceedings/CFE/Xian_2007/2-1_Qili.pdf

The International Iron and Steel Institute. (2005). *Steel: The foundation of a sustainable future: Sustainability report of the world steel industry 2005*. Retrieved September 20, 2011, from http://www.worldsteel.org/pictures/publicationfiles/SR2005.pdf

Kong, Xian. (2002). Developmental direction of energy saving for industrial furnace. *Ye Jin Neng Yuan* [Energy for the Metallurgical Industry], *22*(5), 36–38.

Luo, Dongkun, & Dai, Youjin. (2009). Economic evaluation of coalbed methane production in China. *Energy Policy*, *37*(10), 3883–3889.

Market Avenue. (2008). 2008 report on China's coal industry: Description. Retrieved September 26, 2011, from http://www.marketavenue.cn/Reports_Sample/MAJM041108003.PDF

United Nations Environment Programme. (2006). Energy efficiency guide for industry in Asia: Furnaces and refractories. Retrieved September 26, 2011, from http://www.energyefficiencyasia.org/docs/ee_modules/Chapter%20-%20Furnaces%20and%20Refractories.pdf

United States Energy Information Administration (US EIA). (1999). Natural gas 1998: Issues and trends. Retrieved September 26, 2011, from http://www.eia.doe.gov/pub/oil_gas/natural_gas/analysis_publications/natural_gas_1998_issues_trends/pdf/chapter2.pdf

United States Environmental Protection Agency (US EPA). (2009a). Effluent guideline: Coalbed methane extraction detailed study. Retrieved September 26, 2011, from http://water.epa.gov/scitech/wastetech/guide/cbm_index.cfm#background

United States Environmental Protection Agency (US EPA). (2003). Draft Programmatic Environmental Impact Statement on Mountaintop Mining/Valley Fills in Appalachia. Retrieved September 26, 2011, from http://www.epa.gov/region3/mtntop/eis2003.htm

United States Geological Survey. (n.d.). Coal-bed methane: Potential and concerns. Retrieved September 26, 2011, from http://pubs.usgs.gov/fs/fs123-00/fs123-00.pdf

Wang, Fuchen, & Guo, Xiaolei. (2008). Opposed multi-burner (OMB) gasification technology—New developments and update of applications. Presentation at the Gasification Technology Conference, Washington, DC. Retrieved February 5, 2010, from http://www.gasification.org/Docs/Conferences/2008/37WANG.pdf

World Bank; China Coal Information Institute; Energy Sector Management Assistance Program. (2008). *Economically, socially and environmentally sustainable coal mining sector in China*. Retrieved September 26, 2011, from http://www-wds.worldbank.org/external/default/WDSContentServer/WDSP/IB/2009/01/15/000333037_20090115224330/Rendered/PDF/471310WP0CHA0E1tor0P09839401PUBLIC1.pdf

World Coal Institute. (2009). Improving efficiencies. Retrieved February 5, 2010, from http://www.worldcoal.org/coal-the-environment/coal-use-the-environment/improving-efficiencies/

Xia, Cao. (2006). Regulating land reclamation in developing countries: The case of China. *Land Use Policy*, *24*(2), 472–483.

Xia, Hanping, & Shu, Wensheng. (2003). Vetiver system for land reclamation. Retrieved January 20, 2010, from http://vetiver.org/ICV3-Proceedings/CHN_Land_reclam.pdf

Xiao, Yunhan. (2007). The evolution and future of IGCC, co-production, and CSS in China. Retrieved January 19, 2010, from http://www.iea.org/work/2007/neet_beijing/XiaoYunhan.pdf

Xie, Kechang; Li, Wenying; & Zhao, Wei. (2010). Coal chemical industry and its sustainable development in China. Retrieved January 25, 2010, from http://www.sciencedirect.com

Zhang Cuiqing; Du Minghua; Guo Zhi; & Yu Zhufeng. (2008). Energy saving and emission cutting for new industrial furnace. *Energy of China*, *30*(2008), 17–20.

Energy and Theology

Energy use has been considered a moral and theological issue only since the 1970s. While many religious writers focus on the facts of energy use, many more are reviewing it not only technically but also from political, social, and ethical viewpoints. Most debated is the issue of nuclear energy; currently there is little consensus on its safety or ethical use.

Energy use is both a blessing and a curse. On the one hand, access to energy, especially high-quality energy sources such as fossil fuels and electricity (typically generated from fossil fuels, nuclear fuels, or moving water) enables higher levels of health care, education, and economic success. On the other hand, this same energy use causes environmental and social disruption in the form of mountaintop removal, smog, acid rain, climate change, cancers, and asthma. Yet it was only recently that energy use was recognized as a moral and theological issue. The advent of nuclear power sparked conversation about energy ethics, but it was not until the oil crisis and rising environmentalism of the 1970s that energy significantly entered religious and ethical literature. Religious people have shown their concern about energy use by investigating the technical details of energy production and consumption as well as the political and social ramifications of its use. Additionally, ethicists have developed moral positions on energy use, typically based on existing environmental and social ethics such as concern for the poor and powerless. From this work, many faith-based energy conservation movements have arisen. While religious environmentalists of all faiths are fairly unified in their support of conserving energy, decreasing the use of fossil fuels, and encouraging justice and widespread participation in decision making about energy use, the ethics of nuclear energy are much more contentious.

Religious Education about Energy

Since the mid-1970s popular journals for a religious audience such as *Christian Social Action*, *Christianity and Crisis*, *Church and Society*, *Engage/Social Action*, *National Catholic Reporter*, *Sojourners*, *US Catholic*, and *Witness* have devoted significant space to educating their readers about the basic facts of energy production and consumption. Common topics include statistics about energy use, available energy reserves, and comparisons of energy consumption by country or region. For example, the United States has approximately 5 percent of the world's population but uses about 25 percent of the world's energy. Fossil fuels accounted for approximately 82 percent of the world's marketed energy use as of 2002 and are highly disruptive to the environment and human health. The extraction, processing, and use of fossil fuels cause mountaintop removal, smog, and acid rain, as well as significant increases in rates of asthma, cancer, and other diseases. According to the Intergovernmental Panel on Climate Change, fossil fuel use accounts for 56.6 percent of the anthropogenic greenhouse effect driving climate change (Synthesis Report 2007, 36). Consequently, environmental and social disruption caused by energy use will only increase as climate change worsens. Popularizing facts such as these has been a major part of religious writings on energy; authors want their audience to recognize the negative ramifications of energy use.

Religious ethicists also educate their readers about alternatives to fossil fuels, especially conservation and renewable energy sources including wind and solar. Some also endorse nuclear power, though others vehemently oppose its use due to concerns about safety and justice. Finally, some endorse or critique existing energy legislation, although religious journals typically avoid explicit policy analysis.

Morality and Theology

While educating readers about the technical details of energy use has been a major focus of religious writings on energy, such texts also explore why energy should be a moral issue. To date, religious ethicists have treated energy as a subset of other environmental and social concerns such as biodiversity loss, water pollution, and hunger. Thus, existing theological ethics about energy tend to depend on what preexisting assumptions about the relationship of nature, people and God, gods, or the Ultimate imply for human responsibility—or how the extension of traditional concern for the poor and powerless can be extended to concern for those harmed by environmental degradation.

This article only addresses a few of the reasons why various religious traditions wish to sustain the environment and thus value sustainable energy. Concern for the environment may be rooted in the belief that God created the world good or that the world is ultimately God's, not humanity's. Religious environmental ethics may also be founded on the interdependence of all life, the flow of qi, or the Dao. Belief in sacred lands, the gods in the form of natural entities, and reincarnation may also inform the ecological sensitivities of religious believers. These sorts of beliefs ground the value of humanity and the environment, prompting an energy ethic since energy use can be so helpful and harmful.

Because religious traditions including Buddhism, Christianity, Confucianism, Daoism, Islam, and Judaism prioritize the needs of the poor in their ethics, discussions of energy use rooted in theological ethics may also revolve around concern for the poor and powerless. Thus some religious ethicists focus on the injustice of uneven distribution of affordable access to high-quality energy between countries or among the people within any one country. The use of high-quality energy sources such as electricity, natural gas, and oil enables cooked food and refrigeration, water sanitation, and the basic heating and cooling of homes. When women and children use high-quality energy and no longer have to spend hours a day gathering firewood or trash for fuel, they have more time for education and more time to work for money. Additionally, when high-quality fuels are used, organic matter remains in fields and forests to increase crop yields and decrease environmentally destructive runoff. Thus high-quality energy use enables people to have better nutrition, healthcare, incomes, and educational levels while preserving their local environment.

While increasing energy use can significantly benefit the poor and powerless, they are the first to suffer from the negative side effects of energy use, even if they are not the primary energy consumers. They are more likely to work in or live near environmentally disturbed areas such as mines, refineries, or highways as well as in areas most vulnerable to climate change. Unfortunately, their lack of information, money, time, or political power makes it difficult for many to fight such problems or to choose to work or live elsewhere. Consequently, those who use the least high-quality energy bear the burden of its use; as such justice (or injustice) with respect to the benefits and harms of energy use is a growing focus of religious energy ethics.

Action

For all of these reasons, and many others, religious people encourage action around energy issues. Individual or local activism is most popular and is often encouraged through lists of energy-saving tips: use compact fluorescent light bulbs; weatherize buildings; use energy-efficient appliances; walk or use public transportation; drive an energy-efficient vehicle.

While these activities have been encouraged since the oil crisis of the 1970s, Interfaith Power and Light, started in 1997 as Episcopal Power and Light by Sally Bingham and Steve McAusland, dramatically increased the number of religious organizations conserving energy. Interfaith Power and Light now includes twenty-eight state organizations comprised of over 4,000 congregations representing a cross section of religions in America: Buddhists; Muslims; Jews; evangelical, mainline, and liberal Protestants; and Roman Catholics. Interfaith Power and Light provides moral support, role models, and technical advice to help religious organizations conserve energy, buy renewable-generated electricity, or to advocate for energy and environmental legislation.

While the efforts of individuals or religious organizations to conserve energy are significant, some ethicists emphasize that individual changes are not enough; society as a whole also needs to be transformed. Theological ethicists such as James A. Nash, Ian G. Barbour, and others recognize that laws are necessary to encourage zoning friendly to mass transportation, incentives for conservation, renewable energy development, and recycling, and cap-and-trade systems in which businesses have financial incentives to reduce pollutants. These measures will be necessary to preserve energy and the planet for future generations while meeting the needs of the present.

Nuclear Controversy

While most religious environmentalists are united in their support of conservation and alternatives to fossil fuels, there has been and is much less agreement about the ethics of nuclear energy. Many popular articles illustrate this debate as does *Nuclear Energy and Ethics,* a compilation of views on nuclear energy sponsored by the World Council of Churches. The most common argument against nuclear energy centers on its safety. Radioactive material must be safely mined, refined, manufactured into fuel rods or pellets, and transported to power plants before any energy is harnessed. Power plants themselves must be well designed, maintained, and operated to prevent nuclear plant accidents. Additionally, the spent fuel and other contaminated wastes must be stored for tens of thousands of years to prevent air, water, and land contamination as well as harm to biota, including humans.

These technical challenges cause some ethicists to question whether it is possible to safely utilize nuclear energy and whether it is moral to saddle hundreds of future generations with the responsibility of managing our nuclear waste. The environmentalist and philosopher Kristin Shrader-Frechette observes that Egyptians have been unable to keep their tombs safe for three to four thousand years and thus wonders how humans could keep nuclear waste sites safe for tens of thousands of years amid wars, natural disasters, and social change (Shrader-Frechette 1991, 182).

Others, such as Gordon S. Linsley, a scientist at the International Atomic Energy Agency, argue that there are effective strategies for containing radioactive wastes. Linsley also maintains that the risk of harm from nuclear waste storage is often less than the risks from storing toxic chemicals and heavy metals. If society is willing to accept these risks, he argues, it should also be willing to accept the relatively lower risks of nuclear waste.

The marked difference in the risk assessment of nuclear energy stems in part from a tendency for the public to be more wary of risks that are involuntary, unfamiliar, and catastrophic, as a nuclear accident could be, while technical experts tend to focus on numerical assessments alone. Balancing these two perspectives is an enormous challenge.

The site-selection process for long-term nuclear waste storage facilities also raises moral questions. Should communities accept nuclear waste and the monetary compensation that comes with it to move their communities out of poverty?

The place of nuclear energy in developing countries is also a contentious issue. Some, such as the pathologist Carlos Araoz, the physicist Bena-Silu, and the nuclear physicist B. C. E. Nwosu, see nuclear energy as a sound way to increase the energy use of developing countries and subsequently increase their health, education, and economic power. Others, including environmental epidemiologist Rosalie Bertell and Achilles del Callar strongly oppose nuclear for safety reasons. They think that energy development in developing nations should follow a renewable, conservation-oriented path in order to avoid the problems of energy use in the developed world and the debt necessary for capital-intensive nuclear power. This debate ultimately hinges on the assessment of the risks of nuclear energy and questions of fairness between developing and developed nations, particularly whether the short-term needs of developing countries or the global, long-term environment should be prioritized if the two conflict.

Of course, energy carriers are not chosen in a vacuum; decisions are made between nuclear, fossil fuels, and renewables. In this milieu, nuclear energy gains supporters more concerned about imminent global warming from fossil fuels than long-term nuclear waste storage. Others are unwilling to accept the long-term risks of nuclear wastes. As yet there is no consensus among religious ethicists, environmentalists, or lay people about the safety and morality of nuclear energy, and none is expected soon.

Future Challenges

Despite the widespread controversy over nuclear energy, religious environmentalists generally agree about the necessity of conservation, of moving away from fossil fuels, and of focusing on justice. Future energy sustainability will require more thought about the morality of nuclear energy and the appropriate energy path for developing and developed countries. Attention must also be given to the theological and moral importance of individual *and* societal action. Finally, connecting energy sustainability

to theological issues beyond creation and justice will enable energy sustainability to be thoroughly integrated in the lives and beliefs of religious people.

Sarah E. FREDERICKS
University of North Texas

Further Reading

Araoz, Carlos. (1991). Setting the problem of nuclear energy in the developing world context. In Kristin Shrader-Fréchette, (Ed.), *Nuclear Energy and Ethics* (pp. 72–78). Geneva: WCC Publications.

Barbour, Ian G.; Brooks, Harvey; Lakoff, Sanford; & Opie, John. (1982). *Energy and the American values*. New York: Praeger.

Barbour, Ian G. (1993). *Ethics in an age of technology* (1st ed.). San Francisco: HarperSanFrancisco.

Bena-Silu. (1991). Nuclear technology today: Promises and menaces. In Kristin Shrader-Fréchette, (Ed.), *Nuclear Energy and Ethics* (pp. 55–65). Geneva: WCC Publications.

Bertell, Rosalie. (1991). Ethics of the nuclear option in the1990s. In Kristin Shrader-Fréchette, (Ed.), *Nuclear Energy and Ethics* (pp. 161–181). Geneva: WCC Publications.

Copeland, W. R. (1980). Ethical dimensions of the energy debate: The place of equity. *Soundings, 63*(2), 159–177.

del Callar, Achilles. (1991). The impact and safety of commercial nuclear energy: Perspectives from the Philippines. In Kristin Shrader-Fréchette, (Ed.), *Nuclear Energy and Ethics* (pp. 66–71). Geneva: WCC Publications.

Hessel, Dieter T. (1978). *Energy ethics: A Christian response*. New York: Friendship Press.

Hilton, F. G. Hank. (2001, May 28). Energy and morality 20 years later. *America, 184*(18), 18. Retrieved March 25, 2009, from http://americamagazine.org/content/article.cfm?article_id=930

Intergovernmental Panel on Climate Change. (2007). *Climate change 2007 synthesis report*. Retrieved May 25, 2009 from http://www.ipcc.ch/pdf/assessment-report/ar4/syr/ar4_syr.pdf

Linsley, Gordon. (1991). Radioactive wastes and their disposal. In Kristin Shrader-Fréchette, (Ed.), *Nuclear Energy and Ethics* (pp. 29–54). Geneva: WCC Publications.

Nwosu, B. C. E. (1991). Issues and experiences concerning nuclear energy and nuclear proliferation. In Kristin Shrader-Fréchette, (Ed.), *Nuclear Energy and Ethics* (pp. 79–88). Geneva: WCC Publications.

O'Neill, John, & Mariotte, Michael. (2006, June 16). Nuclear power: Promise or peril. *National Catholic Reporter*, 17–18.

Shrader-Fréchette, Kristin. (Ed.). (1991). *Nuclear energy and ethics*. Geneva: WCC Publications.

Smil, Vaclav. (2003). *Energy at the crossroads: Global perspectives and uncertainties*. Cambridge, MA: MIT Press.

Tester, Jefferson W.; Drake, Elisabeth M.; Driscoll, Michael J.; Golay, Michael W.; & Peters, William A. (2005). *Sustainable energy: Choosing among energy options*. Cambridge, MA: MIT Press.

United Nations Development Programme. (2005). *Energizing the millennium development goals*. New York.

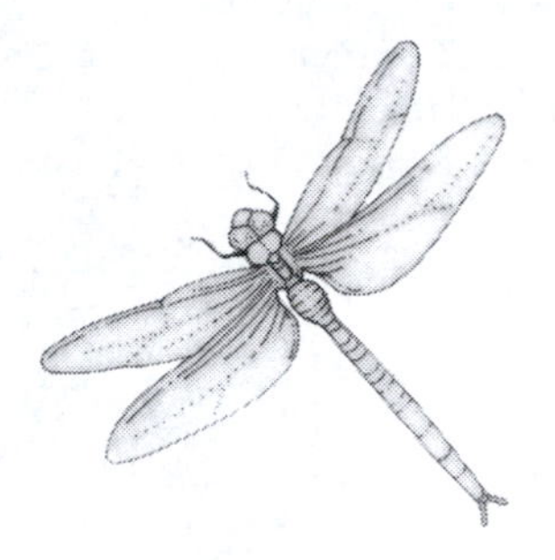

Energy Conservation Incentives

Energy conservation and energy efficiency became a public policy issue in the United States, Europe, and Asia during the Middle East oil embargo of the 1970s. Legislation in the United States was enacted to promote conservation among consumers and industry, but within several years, the emphasis turned to energy efficiency. Through the years, legislation has promoted various initiatives for consumers including tax benefits and rebate programs.

Environmentalists, the media, and the general pubic, have all intermittently used the term *energy conservation* to define the idea of both "conservation" (i.e., using less energy) and "efficiency" (i.e., getting more out of the same energy source). Although traditional conservation has been the subject of numerous laws and policies in the United States and abroad, governments have tended to focus on reducing energy use through measures that encourage efficiency. Energy efficiency policies are easier to implement and technologically more feasible than those that focus solely on conservation, explaining why government intervention has leaned in this direction.

Prior to the 1970s, most of the industrialized world did not much consider the amount of energy they used or the price they paid for it. This was especially true in the United States, which had recently undergone a period of unprecedented prosperity. Americans were enamored with new household items such as televisions and dishwashers. These items were sold as the epitome of modern convenience, and no one seemed to worry about the amount of energy they consumed. But the situation soon changed.

In 1973, Arab oil-exporting countries declared an oil embargo against the United States, Canada, western Europe, and Japan. The embargo was spurred by the United States' decision to supply the Israeli military with arms during the Yom Kippur War between Syria and Egypt. Access to oil was severely diminished, and hence the price rose dramatically. The embargo had a widespread effect on the economies of the United States and Europe. Prices of retail gasoline rose in the United States from a national average of 38.5 cents in May of 1973 to 55.1 cents in June of 1974. State and federal governments called for conservation by asking businesses and individuals to alter their energy habits. This was the first time in history that the United States was not in control of their energy supply, and conservation was the call to arms (Kelley 2008).

The oil embargo altered the energy landscape, not only for the United States, but for the world. It fostered a continuing debate about how to reduce our use of energy, whether through government mandates, voluntary programs, utility regulation, or all of the above. Although oil prices have stabilized, the environmental impact of energy production has remained an important topic. Conservation and energy efficiency are once again being considered as the means to reduce energy use and, therefore, demand. The question still lingers as to whether market signals alone enable consumers to make rational energy decisions, or whether tax relief or other incentives are needed to push them in the direction of efficient energy choices.

Conservation through Legislation

During the 1970s, the impetus behind all action to reduce our energy use was the price of oil and energy security. In the United States, the federal government began passing legislation to deal with the oil crisis, including the Weatherization Assistance Program (WAP) of 1976 to help low-income customers reduce their energy use through insulation. The Low-Income Energy Assistance Program (LEAP), run by the Office of Health and Human Services, was also started to help low-income customers deal with the rising price of energy. Both of these programs remain in existence. In 1978, Congress passed the National Energy Conservation Policy Act (NECPA), which required utilities to offer energy efficiency and demand-side management programs to their customers. For the first time, customers were offered incentives including low-cost loans and cash to replace inefficient appliances. Utility companies began other programs such as time-of-use rates and load shifting. It was unclear at the time whether these laws and incentives would affect customers' energy choices and have a real impact on energy demand and prices, but most industry experts believed that the only viable method to promote energy conservation was through government regulation.

Views toward energy conservation and efficiency changed in the 1980s as oil prices came down and energy prices stabilized. Politics now focused on market-oriented polices as a means of altering behavior (Sanstad and Howarth 1994). Energy price signals alone were no longer driving customers to conserve energy. At the same time, the environmental impact of energy use began to gain attention as a significant issue. Questions arose as to whether or not market forces were sufficient to incentivize energy efficiency and conservation to achieve environmental goals (Kelley 2008). Even though over thirty years of research has been conducted regarding what drives consumers to make their energy decisions, this debate continues today (Sanstad and Howarth 1994).

Bounded Rationality

In a perfect world, energy prices would reflect all externalities, including private and social costs of production, and consumers would make purely rational energy choices based on a cost benefit analysis. This, of course, is not reality. Energy prices do not include environmental consequences, and rarely do they include transaction costs (Sanstad and Howarth 1994). Therefore, consumers have imperfect information and do not perform the cost benefit analysis required to make purely rational energy choices. They would require real time pricing and would consider all externalities when choosing appliances, heating units, or housing. Consumers rarely consider energy usage or efficiency when making these decisions, although in the past ten years, environmental costs have played an increasingly important role. This phenomenon is known as "bounded rationality" (Sanstad and Howarth, 1994).

Research has shown that the information presented to consumers is not only incomplete but also incorrect (Stern 1986). Even though energy users may be motivated to make good choices—ones that take into account environmental harm for instance—they are not able to do so based on pricing alone. Behavioral research also has shown that consumers deviate from "preference maximization," meaning they lack the expertise to carry out the numerical cost benefit analysis associated with their actions (Sanstad and Howarth 1994). This means there are two forces at work: (1) energy prices are on their face incorrect; and (2) consumers are unable to make purely rational energy choices. These two facts are relied upon as justification for energy policies and programs that promote energy efficiency and/or conservation. The debate then focuses on what types of incentives are needed: government mandated policies, voluntary programs, or a combination of these factors. What types of policies and programs are more successful in lowering energy usage, those that focus on behavior or those that focus on energy intensive industries?

Intervention

Once it is assumed that some type of intervention is needed to encourage energy efficiency and conservation in order to slow down the environmental impact of energy use, the question becomes what types of incentives are needed and who should be receiving them. Do these incentives actually make a difference? Since we know that energy prices do not reflect environmental consequences, we must then consider how to (1) encourage the production of more efficient technology, and/or (2) encourage consumers to choose the more efficient appliance, car, or house, assuming it is available, and/or (3) encourage the consumer to use less energy themselves.

Utility regulation was an early method relied upon by government to incentivize energy conservation during the days of the oil crisis. Utilities were required to offer programs that permitted customers to take charge of their demand and offer different rates depending on the time of day a customer used energy. Utility programs have had a limited effect, however, due to their heavy reliance on the consumer. It is believed that when consumers are left to their own devices, they are unable to make choices that significantly impact energy use (Nadel 1990).

Cost subsidies are another type of incentive and are quite easy to implement. It is an umbrella term that can include incentives ranging from tax benefits to rebate programs. The Energy Tax Act of 1978 was the first legislation in the United States that provided a tax credit to consumers for investing in energy conservation. More recently, the American Recovery and Reinvestment Act (2009) provided over $20 billion in funds to promote energy efficiency investments, and $5 million was specifically earmarked to fund the Low-Income Weatherization Assistance Program. Energy experts believe that tax credits for such measures as adding insulation or replacing inefficient home appliances have a positive effect on lowering energy demand (Hasett and Metcalf 1995). Although access to funds does provide consumers with some incentives to make investments in energy efficiency (i.e., rebates and tax credits for replacing appliances or for weatherization of the home), it is still unclear whether these investments will conserve energy on any significant level.

Both federal and state governments originally focused on utility regulation as a means of reducing energy consumption, but mandatory efficiency standards for appliances and building equipment have had the most success in reducing energy use over the long-term. In 1987, President Ronald Reagan signed the National Appliance Energy Conservation Act (NAECA). NAECA was intended to be updated periodically to deal with technological advances in efficiency as they became available (Kelley 2008). Four presidents since have amended the legislation to add equipment and upgrade the standards. It is believed that these standards have avoided the building of new power plants and transmission lines in the United States. The American Council for an Energy Efficient Economy (ACEEE) estimates that the cumulative energy savings derived from these standards during the years 1990–2030 will save consumers in excess of $180 billion (Kelley 2008). These standards have even greater potential for creating energy savings in the future as old equipment is retired and improvements in technology are made.

The Energy Star program is another successful method for reducing energy use via energy efficiency. Energy Star is a voluntary US federal program introduced in 1992 by the Environmental Protection Agency. It provides a seal of approval for everything ranging from dishwashers to new homes that meet its high energy-efficiency standards. The Energy Star standards exceed those mandated by federal legislation for appliances and building equipment and also include items that are not required to meet the federal standards. Moreover, in 1996, Energy Star teamed up with the Department of Energy to expand its reach, and it now includes building construction methods (Kelley 2008). Energy Star also is working with the European Union to coordinate efficiency standards for office equipment due to the global market for such products. This voluntary program has proven to be an effective way to allow consumers to take energy into consideration when making purchasing decisions.

The success of both the appliance and equipment mandates, as well as the Energy Star program, demonstrate that the preferable method of reducing energy use is by focusing on technology as opposed to focusing on consumer behavior. Intervention is more productive when the policies meant to promote conservation and efficiency do not rely solely on the choices of consumers (Nadel 1990). Even though consumers may want to consider the externalities of their energy choices, they may not have the information or the expertise to do so. Therefore, the policies that will have the most success in promoting energy conservation will be those that reduce the amount of energy used by energy intensive technologies and products.

Adrian DiCianno NEWALL
Energy Consultant

See also in the *Berkshire Encyclopedia of Sustainability* Energy Subsidies; Environmental Law—Arab Region; Environmental Law—United States and Canada; Green Taxes; Investment Law, Energy; Utilities Regulation

Further Readings

Haslett, Kevin A., & Metcalf, Gilbert E. (1995). Energy tax credits and residential conservation investment. *Journal of Public Economics, 57*(2), 201–217.

Kelley, Ingrid. (2008). *Energy in America: A tour of fossil fuel culture and beyond.* Burlington: University of Vermont Press.

Lutzenhiser, Loren. (1993). Social and behavioral aspects of energy use. *Annual Review of Energy and the Environment, 18*, 247–289.

Nadel, Steven. (1990). Electric utility conservation programs: A review of the lessons taught by a decade of program experience. *Proceedings of the ACEEE 1990 Summer Study on Energy Efficiency in Buildings* (Volume 8). Washington, DC: Americian Council for an Energy-Efficient Economy.

Sanstad, Alan H., & Howarth, Richard B. (1994). Consumer rationality and energy efficiency. *Proceedings of the ACEEE 1994 Summer Study on Energy Efficiency in Buildings.* Washington, DC: American Council for an Energy-Efficient Economy.

Stern, Paul C. (1986). Blind spots in policy analysis: What economics doesn't say about energy use. *Journal of Policy Analysis and Management, 5*(2), 200–227.

Energy Efficiency Measurement

Energy efficiency is easy to define but hard to measure. Indicators such as energy intensity are used as proxies for energy efficiency, but their results can be misleading. Efficiency improvements, mainly caused by stock turnover to more efficient products, are a normal by-product of economic growth and modernization. Government policies to promote energy efficiency have not led to reductions in national energy use, mainly due to rebound effects.

Energy efficiency is easy to define in theory but hard to measure in practice. At its simplest, it is defined as the ratio of the energy output to the energy input for a machine. For a boiler or an engine, the energy input (the fuel) and the output (the heat or power produced) thus can be measured fairly easily. For machines (or other devices) where the output is not simply energy but some form of energy service, however, measuring efficiency is more complex. For instance with a lightbulb, we are interested not in its heat output but its light output. The same with a refrigerator, an air conditioner, a car, or a computer—we are interested in the machine's ability to perform an energy service by cooling, transporting, or information processing. Evaluating the energy efficiency depends on defining the desired output: what aspect of the energy service are we really interested in—power, speed, reliability, or economy?

The more complex the energy system, the harder it is to agree on an output measure. How do you measure the output of a bakery, a restaurant, a hotel, a chemical factory, or an office? One approach has been to express energy efficiency in terms of energy intensity: energy input per unit of some (measurable) factor, such as physical or monetary output, number of workers, or floor area. When it comes to comparing industries, transport systems, or countries, the problems are immense. Measuring energy inputs is fairly easy—there are good energy statistics on national and sectorial energy production and on imports and exports—but what is the most suitable measure of industrial or national output? Is it some measure of monetary output or gross domestic product (GDP)? If so, we can then reach some trivial conclusions based on energy intensity: an investment firm's office is more efficient than a steel works; Denmark is a more efficient country than the United States. As a review of energy efficiency polices in International Energy Agency (IEA) countries admitted, "Trends in final energy intensity are often used to assess the extent to which energy efficiency is improving in countries. However, this can often be misleading as the ratio is affected by many factors such as climate, geography, travel distance, home size and manufacturing structure" (Taylor et al. 2010, 6468).

If our measure of transport efficiency is kilometers per liter, then a car is more efficient than a bus. If we account for the number of passengers, by using an index of passenger-kilometers per liter, this gives a better indication of efficiency; however, the efficiency measure now depends on the number of passengers on the bus (which depends on the time of day and route) and how often it stops. Similarly, trying to determine if it is more efficient to drive or fly requires making many assumptions, such as how full the car and plane are, the distance traveled, and even how people travel to and from the airport. Overall, however, a full plane or train is generally more efficient than a single-user car.

The challenge of trying to construct an energy efficiency index is that in a complex energy system there are so many variables that it is not possible to account for them all. For instance, a house can have a very efficient boiler but be ineffective overall in providing the standard of heating required by the occupants. This could be because the house is poorly insulated, so that much of the (efficient) heat produced by the boiler is lost through the walls, roof, and windows, or because there are few heating controls, so unoccupied rooms are heated. Similarly, a factory having very efficient equipment could be judged as inefficient overall if it is running at low capacity or produces poor-quality goods that fail to sell. It is not the owning of efficient machines that makes a factory (or a country) efficient, but how those machines are used. For example, exporting the most efficient machinery to developing countries is no guarantee of efficient industries or production.

Industrialization, through the use of modern energy efficient equipment, will undoubtedly lead to lower energy intensity, as it has done in China. But as the economy expands, energy consumption grows. This has only been considered a problem if energy consumption causes major environmental problems, such as local smoke or air pollution. Improved energy efficiency equipment can reduce these local problems, such as the replacement of open coal fires with gas central heating.

Technical Innovation

Changes in energy efficiency are an inevitable consequence of technical innovation and of manufacturers' desires to reduce costs of production through reducing material and energy inputs. Consumer choice of an energy-using product is seldom based on one attribute, such as energy efficiency; rather it is a compromise between price, performance, and operating costs. Designations of a product as "energy efficient," like labels such as "green," "powerful," or "glamorous," are advertising claims based on comparative measures. A modern compact fluorescent lightbulb (CFL) may be considered more energy efficient than the old incandescent bulb (in terms of light output per unit of electricity input), but they are not identical products. There are differences in purchase cost and lifetime, as well as consumer concerns about light color, performance, and suitability for certain fixtures. A subcompact car may be more efficient (in terms of kilometers per liter) than a large luxury sedan, but they are not similar vehicles in terms of size, power, or features. Technical innovation can produce more efficient products, but consumers will only buy them if they rate energy efficiency high among desirable attributes. Driven by the desire to cut (operating) costs or the perceived need to reduce energy use (due to national or global concerns), some consumers do place a high value on energy efficiency.

Environmental Interest

Until the mid-1970s, energy efficiency was a concept confined to the world of industrial engineers and factory managers and was closely allied to productivity concerns. It was then adopted by the emerging environmental movement and was heavily promoted by Amory Lovins in his book *Soft Energy Paths* (1977) as the alternative solution to nuclear power (and challenged by the British energy expert Len Brookes; see, for example, Brookes 2000). Lovins's argument was that consumers, by using more energy efficient equipment could get the same energy service for less fuel input, thus reducing national energy consumption (and the need to build new power plants). This new, or "soft," energy path, would involve the use of CFLs, condensing boilers, A+ rated refrigerators, and improved building insulation.

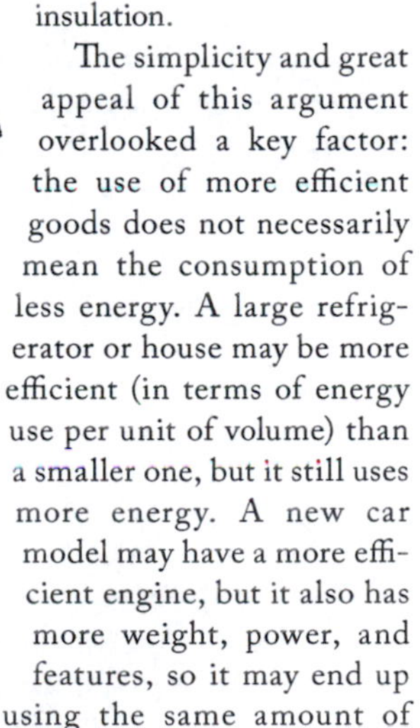

The simplicity and great appeal of this argument overlooked a key factor: the use of more efficient goods does not necessarily mean the consumption of less energy. A large refrigerator or house may be more efficient (in terms of energy use per unit of volume) than a smaller one, but it still uses more energy. A new car model may have a more efficient engine, but it also has more weight, power, and features, so it may end up using the same amount of fuel. A new house may have more insulation than an older one, but it is larger in area, and it has air conditioning. Furthermore, manufacturers and utilities have historically used increased energy efficiency to promote increased consumption. More energy efficient products and practices does not equate directly to lower (national) energy consumption, especially if the economy and the population are growing.

Nevertheless, environmentalists, followed by many governments and industry groups, promoted the efficiency message as a relatively painless and cost-effective way to solve the problems of energy supply, and, by extension, as the ideal solution to the problems of global warming. This argument rests on the belief that global or national energy use can be reduced by the sum of the millions of small energy savings by consumers through their adoption of energy efficient devices—that is, what is true at the micro (local) level is also true at the macro (national or global) level. At the micro level, energy efficiency does save energy (and the monetary savings are often used to cover the cost of investment), but the key question is whether these micro-level savings are cumulative or whether they get (re)spent and ultimately dissipated. Is energy saved never used, or is it merely diverted to another use? This is the rebound effect question.

Rebound Effect

Rebound effects, first identified by the nineteenth-century British economist Stanley Jevons, are normal occurrences in economic systems: as improvements in efficiency, due to technical change, make goods and services cheaper, we can and generally do buy more of them. If the cost of an energy service, such as heating, falls, we can afford to use more, by heating our rooms to higher temperatures, or for longer periods, or heating more of the house (a direct rebound effect). We could also use the money saved to buy other goods and services that use energy (an indirect rebound effect). For producers, more efficient production processes (e.g., in a steel industry) result in lower cost both for that commodity (cheaper steel) and for products made from that commodity (e.g., cars). Cheaper steel results in greater sales and hence more production; cheaper cars result in more car sales and increased car travel. The end result is greater energy use in the long run that can outweigh the original efficiency saving.

When we save energy through improved efficiency but then consume only part of the saved energy, that is a rebound of less than 100 percent. If, however, that efficiency encourages us to consume more energy than we would have, that rebound of more than 100 percent is sometimes called a backfire effect. The rebound effect is linked to the major policy question: if energy use continues to grow despite large increases in energy efficiency, then is promoting greater energy efficiency a credible government policy to reduce national energy use, and hence carbon emissions?

Government Policy

The role governments should play in promoting energy efficiency is not at all clear. If the desire is to reduce energy use in the short term, perhaps due to some national crisis (such as the shortage of electricity in Japan after the devastating earthquake in March 2011), then regulations and social incentives can be used to change the behavior of consumers. Japan has a long history and culture of promoting energy conservation (or frugality), employing the cutting back and careful use of energy consumption as energy conservation measures. This was reinforced in 1979 by passage of the Law Concerning the Rational Use of Energy, which encouraged the development of more energy efficient appliances and industrial products and processes, with the overall aim of making Japan the world's most energy efficient nation.

Many Western governments have tried to influence their consumers to purchase more efficient products, through appliance subsidies and building regulations. It is not clear, however, whether this is good use of public money. Might people have bought the more efficient appliance without a subsidy? Should old, poorly insulated buildings be refurbished, or is it "better" to demolish them and rebuild with new, efficient buildings?

Energy efficiency policy, like all policy areas concerned with changing people's choices, is full of social and economic conflicts and can lead to unintended consequences. Chief among these is the rebound effect: greater efficiency lowers operating costs and opens up the market to more consumers. This is particularly the case where there is a large unmet demand for an expensive energy service (such as for air-conditioning, heating, or lighting) or where a technology or product is, or could be, widely used in industry or commerce (such as computers, electric motors, cement, or steel, so-called general purpose technologies).

The Chinese government set a goal in November 2005 to reduce its energy intensity (energy use per unit of GDP) by 20 percent between 2006 and 2010, a seemingly modest goal given that energy intensity had already decreased 5 percent annually between 1970 and 2001. This decline in intensity was due to the very rapid economic growth during these decades, which resulted in large increases in energy consumption (with the side effect that China overtook the United States to become the world's leading emitter of carbon dioxide in 2006). China's goal of improving its energy efficiency (as measured by declines in energy intensity) rests on a complex array of fiscal, pricing, and technical measures that affect all sectors of the economy. The most important is the Top-1000 Energy-Consuming Enterprises program, which targets the 1,000 largest industrial enterprises, accounting for a third of national energy consumption.

The goal of most national energy efficiency programs, however, is not to reduce absolute energy use or carbon emissions. Rather the goal is to reduce energy consumption below a projected forecast—that is, what the energy consumption would have been without the efficiency programs. The stated savings of most energy efficiency programs thus are hypothetical; energy use (at the macro level) usually increases even though the chosen indicators of energy efficiency, such as energy intensity, may show the desired decline.

Energy Sufficiency and Conservation

An alternative strategy to energy efficiency is energy conservation that seeks to cut energy consumption by decreasing levels of energy service, mainly through changes in behavior and lifestyle. This includes things like reducing heating levels, driving slower, using smaller appliances, and generally consuming less.

This strategy of consuming less is called "sufficiency," or in the United States, "simple living," "voluntary simplicity," or more generally, "downsizing," often done (voluntarily) through deciding to work and earn less. So how feasible is it to expand the idea of sufficiency to a national scale, by cutting consumption of energy-intensive goods and services like frozen foods or air travel? Furthermore, would the widespread adoption of sufficiency have any impact on global resource or energy use? Once again the rebound effect has an impact, and the widespread impact of reducing demand for a commodity (like energy) results in lowering its global price and allowing greater consumption by marginal consumers.

So would our Western reduction of consumption merely allow poorer consumers elsewhere to enjoy a higher standard of living? Given the lower environmental standards in developing countries, a shift in energy demand from rich to poor countries could result in greater environmental pollution. If we use less electricity from our clean coal power stations, will China and India use more in their dirty ones? This is a complex debate that has only just started.

Outlook

Under conditions of economic growth, national energy use has historically increased despite big improvements in energy efficiency. Past attempts to promote energy efficiency as a means of reducing national energy consumption have proved futile. It has even been argued that promoting energy efficiency (like any other factor of production such as labor or resources) stimulates economic growth, and hence energy use. While improved energy efficiency, mainly brought about by stock turnover (the replacement of old, inefficient equipment by new, more efficient equipment) is good for the economy and for people's lives, it is unlikely to be a solution to the problems of energy shortages or climate change. Energy efficiency makes us richer, however, through lowering operating costs, and allows us (if we so choose) to invest our money savings in new or low-carbon energy sources. Energy efficiency is a means to a sustainable world, not an end in itself.

Horace HERRING
The Open University

See also in the *Berkshire Encyclopedia of Sustainability* Building Rating Systems, Green; Carbon Footprint; Computer Modeling; Development Indicators; Energy Labeling; Human Appropriation of Net Primary Production (HANPP); I = P × A × T Equation; International Organization for Standardization (ISO); *The Limits to Growth;* Material Flow Analysis (MFA); Shipping and Freight Indicators; Social Life Cycle Assessment (S-LCA); Supply Chain Analysis

Further Readings

Andrews-Speed, Philip. (2009). China's ongoing energy efficiency drive: Origins, progress and prospects. *Energy Policy, 37*(4), 1331–1344.

Balachandra, P.; Ravindranath, Darshini; & Ravindranath, N. H. (2010). Energy efficiency in India: Assessing the policy regimes and their impacts. *Energy Policy, 38*(11), 6428–6438.

Brookes, Len. (2000). Energy efficiency fallacies revisited. *Energy Policy, 28*(6–7), 355–366.

Calwell, Chris. (2010, March 22). Is efficient sufficient? The case for shifting our emphasis in energy specifications to progressive efficiency and sufficiency. Retrieved September 18, 2011, from http://www.eceee.org/sufficiency

Fouquet, Roger. (2008). *Heat, power and light.* London: Edward Elgar.

Herring, Horace, & Sorrell, Steve (Eds.). (2009). *Energy efficiency and sustainable consumption: The rebound effect.* Basingstoke, UK: Palgrave.

Jaccard, Mark. (2005). *Sustainable fossil fuels.* Cambridge, UK: Cambridge University Press.

Jackson, Tim. (Ed.). (2006). *The Earthscan reader in sustainable consumption.* London: Earthscan.

Jenkins, Jesse; Nordhaus, Ted; & Schellenberger, Michael. (2011). *Energy demand: Rebound and backfire as emergent phenomena.* Retrieved September 18, 2011, from http://thebreakthrough.org/blog/2011/02/new_report_how_efficiency_can.shtml

Lovins, Amory. (1977). *Soft energy paths.* London: Pelican.

Polimeni, John; Mayumi, Kozo; Giampietro, Mario; & Alcott, Blake. (2008). *Jevons' paradox and the myth of resource efficiency improvements.* London: Earthscan.

Smil, Vaclav. (2003). *Energy at the crossroads.* Cambridge, MA: MIT Press.

Stewart, Devin, & Wilczewski, Warren. (2009, February 3). How Japan became an efficiency superpower. Retrieved September 18, 2011, from http://www.policyinnovations.org/ideas/briefings/data/000102.

Taylor, Peter; d'Ortigue, Oliver; Francoeur, Michel; & Trudeau, Nathalie. (2010). Final energy use in IEA countries: The role of energy efficiency. *Energy Policy, 38*(12), 6463–6474.

Weizsacker, Ernst; Hargroves, Karlson; Smith, Michael; & Desha, Cheryl. (2009). *Factor five: Transforming the global economy through 80% improvements in resource productivity.* London: Earthscan.

Zhou, Nan; Levine, Mark; & Price, Lynn. (2010). Overview of current energy-efficiency policies in China. *Energy Policy, 38*(11), 6439–6452.

Energy Industries—Bioenergy

Derived from natural resources, bioenergy is a prominent renewable energy source throughout the world. Most simplistically, it is wood used for cooking and heating; more complex examples include thermochemical conversions that produce biofuels. Although relevant technologies exist (and continue to be developed), many are not yet cost-effective, and questions remain about the industry's long-term environmental impact.

Bioenergy is any energy, or fuel used to produce energy, that is derived from either biomass or plant matter. According to the US Energy Information Administration (2009), the United States consumed more bioenergy in 2008 than wind, solar, and geothermal energies combined. Despite its widespread use around the world, the term is often mistakenly equated with liquid biofuels such as ethanol and biodiesel. Biofuels—solid, liquid, or gaseous—are virtually the only liquid fuel alternative to oil for transportation purposes; they can be used not only for fueling vehicles but also for heating, cooling, and electrical power generation.

In many parts of the world, bioenergy remains the dominant energy source for heating and cooking, but bioenergy is much more than wood used in the home or liquid biofuels like ethanol. Bioenergy production includes biogas and electricity generation from such sources as food-processing wastes and the anaerobic digestion of cow manure. (Anaerobic bacteria digest matter in the absence of oxygen, yielding a biogas rich in methane and carbon dioxide.) Bioenergy is being used in district heating systems; for example in St. Paul, Minnesota, water heated from waste wood heats the central business district of the city.

Interest in bioenergy is growing for several reasons. Economic developers and rural communities see it as a growth industry that can create jobs and revive local economies. Proponents of agriculture and forestry see it as a tool to protect productive working landscapes and provide new markets for these volatile industries. Environmentalists see bioenergy as a means to reduce greenhouse gas emissions associated with fossil fuel use. Against this backdrop, policy makers are looking for ways to guarantee that rural communities capture the full benefits of sustainable bioenergy production (Radloff and Turnquist 2009).

Bioenergy as a Renewable Resource

Energy derived from biomass is considered renewable because bioenergy is simply stored solar energy from the sun. It can be produced from a wide range of biomass types including plants, animals, and animal wastes. Biomass includes conventional crops such as corn and soybeans, vegetable oils, agricultural residues like corn stover and rice straw, wood and forest residues, and mill residues. Other types of biomass, also referred to as feedstock, include construction waste, perennial grasses like switch grass, and short rotation woody crops (including willow and hybrid poplar). Animal renderings and animal manures are additional examples of biomass (Biomass Research and Development Board 2008).

Because bioenergy is not classified in a consistent manner, this article identifies three broad "types" of bioenergy based on *how* the energy derived from biomass is used: bioheating/biocooling, biopower, and biofuels.

Bioheating/Biocooling

Bioheating/biocooling is the most prevalent use of biomass worldwide, especially in parts of the developing world where it is the primary energy source for cooking. In countries like Nepal, Sudan, and Tanzania, about 80–90 percent of all energy comes from biomass (Rosillo-Calle et al. 2007). In the United States, the residential home heating market is the largest user of wood fuel. Other bioheating/biocooling examples include schools and commercial buildings that heat and cool with wood, as well as district heating systems that provide heating and cooling to multiple buildings through a system of connected pipes carrying hot water or steam.

Biopower

Biopower is electricity produced from biomass, and it is often generated through cofiring of biomass with coal. Heat from the combustion process drives a steam turbine that generates power. Alternative methods of generating biopower include gasification, a technology that converts biomass under high temperatures into a gas called "syngas." The syngas can then be converted to electricity using a more efficient technology than a conventional steam turbine. Several additional biopower technologies exist. Combined heat and power (CHP) technology increases energy efficiency by generating electricity and capturing and using any excess heat produced in the process.

Biofuels

The third category of bioenergy is biofuels, which include liquids as well as solid. Conventional liquid biofuels (like ethanol) come from existing food crops. More advanced liquid biofuels may come from cellulosic sources such as wood and perennial grasses. Solid biofuels (including wood pellets) are currently derived from wood waste and mill residues. As the demand for bioenergy grows, more solid biofuels may be manufactured from additional sources including perennial grasses, construction waste, and other waste streams.

Bioenergy is unlike other renewable energy resources in two key aspects. First, bioenergy is not an intermittent source of energy. Unlike wind and solar, which depend on windy and sunny days respectively to generate power, bioenergy is an on-demand source of renewable energy that can be dispatched where needed. As long as the feedstock supply is adequate, bioenergy represents a dependable, constant source of energy that can be generated at any time of the day as needed—independent of daily weather patterns.

Second, biomass as an energy source is a complex system of interdependent components. According to the US Environmental Protection Agency (US EPA 2007), economically and technically feasible bioenergy projects require an adequate feedstock supply, effective conversion technologies, dependable markets, and viable distribution systems. The challenge of ensuring an adequate feedstock supply is particularly challenging.

Converting biomass into energy may be a complex process, but it can be broken down into three major phases: growing and transporting feedstocks, converting feedstocks into bioenergy, and marketing bioenergy.

The Biomass Supply Chain

Growing and transporting feedstocks is known as the biomass supply chain. The first step in the process is the growing and/or harvesting of available biomass and nonagricultural waste. Biomass feedstocks can be lumped into three general categories: conventional feedstocks, dedicated feedstocks, and waste or underutilized feedstocks.

Conventional feedstocks include wood waste, grains, and other common forms of biomass. These types are easy to convert to bioenergy because systems are in place to grow them, harvest them, and process them using existing technologies. Examples include corn grain (for ethanol) and trees (for making pellets or for combusting directly).

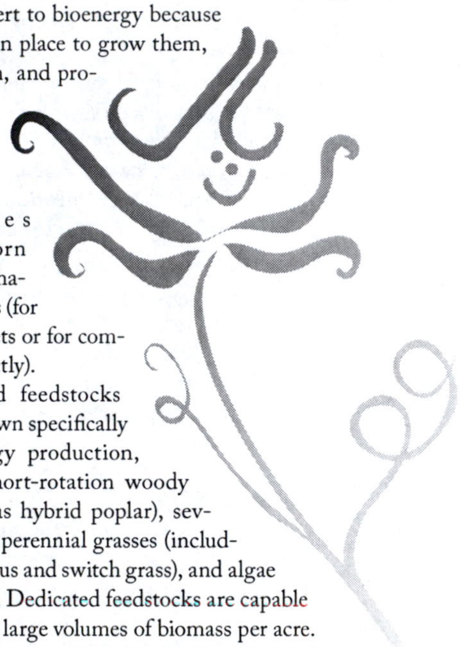

Dedicated feedstocks are crops grown specifically for bioenergy production, including short-rotation woody trees (such as hybrid poplar), several types of perennial grasses (including miscanthus and switch grass), and algae and jatropha. Dedicated feedstocks are capable of producing large volumes of biomass per acre.

Waste or underutilized feedstocks refers to those types of biomass commonly thought of as waste products like food-processing waste, brush, tree tops and trimmings, construction and demolition debris, leaves and yard waste, animal renderings, and manure. As some of these feedstocks become increasingly valuable, it's clear they are not really "wastes" at all; rather they are historically underutilized waste products that in many cases make good bioenergy feedstocks. The second step in the biomass supply chain is to process the available biomass into a feedstock. For example, slash from a recent timber harvest may need to be bundled prior to transport. The feedstock is then aggregated and delivered to some type of biorefinery for processing.

Converting Biomass Feedstocks into Energy

The next step in the process of bioenergy production is the conversion of the feedstock into intermediate products such as combustible gases, carbon dioxide, oils, tars, and liquids. These intermediate products are then converted into final, usable energy products such as electricity, heat, and solid and liquid fuels using one of five basic conversion technologies, ranging from simple to more complex.

1. **Physical conversion.** This is the simplest way to convert biomass into a usable energy form. Making wood pellets is one example, and straight vegetable oil (SVO) production is another. With SVO production, oil is simply extracted through a seed-pressing process. Pressed vegetable oil can be used as a transportation fuel in tractors and in diesel engines. In both the wood pellet and vegetable oil examples, the biomass feedstock is physically altered through force, producing a fuel that can be used for transportation or converted into another energy product through one of the other conversion technologies discussed below.
2. **Combustion.** Combustion technologies convert biomass into hot air, hot water, and steam. They range from smaller household technologies like wood stoves to large commercial and industrial technologies including fixed bed combustion and fluidized bed combustion systems. Larger commercial and industrial combustion systems rely on wood chips, corn stover, bark, and other less processed feedstock, while smaller household-size technologies require higher quality fuels.
3. **Chemical conversion.** Through chemical conversion, feedstocks are decomposed into liquid biofuels (US EPA 2007). An example of chemical conversion is transesterification, the process used to make biodiesel in which oils, fats, used cooking greases, and other fatty wastes are combined with a catalyst such as methanol. The final products are biodiesel and glycerin, which is often used in soaps.
4. **Biochemical conversion.** Through biochemical conversion, enzymes and bacteria break down feedstocks like cow manure and perennial grasses into intermediate products such as biogas. Biogas is similar to natural gas and contains impurities including sulfur, carbon dioxide, nitrogen, and hydrogen, but biogas can be cleaned up and used in a similar fashion as natural gas. It can be combusted to generate electricity or compressed for other uses, including transportation. Examples of biochemical conversion technologies include simple composting, bioreactors at landfills, and anaerobic digesters on farms and at wastewater treatment facilities.
5. **Thermochemical conversion.** The thermochemical conversion method is similar to the biochemical conversion method in that it produces intermediate products that are then further refined into useful end products. With thermochemical conversion, however, the biomass feedstock is decomposed through the use of heat instead of with enzymes and bacteria. Intermediate products resulting from thermochemical conversion of biomass include combustible gases, liquids, tars, and charcoal. These products can be further refined into many different final products: ethanol, diesel, gasoline, hydrogen, and bio-oil. Examples of thermochemical technologies include gasification and pyrolysis (the chemical decomposition of organic matter using heat).

Bioenergy Opportunities

Making bioenergy is relatively easy; making a profitable business out of bioenergy production is not. In order for the bioenergy industry to thrive, it needs a reliable market. Bioenergy can either be used on-site, in the local community, or exported to another region. Using bioenergy on-site is an attractive option for many energy producers, because on-site use can directly substitute for off-site energy purchased at retail prices. Selling bioenergy locally is another attractive option because it cuts down on transportation costs.

There are several social, economic, environmental, and technical opportunities associated with greater bioenergy use. In rural communities, increased use of bioenergy may stimulate the economy and create cooperative and local ownership opportunities related to the growing, transporting, refining, and marketing of biomass. Enhanced bioenergy use could also make biomass-rich communities more energy independent and less susceptible to fluctuating energy prices. While expanding social and economic opportunities, bioenergy could also help the environment. Bioenergy can be produced from feedstocks such as perennial grasses, which sequester carbon underground through their vast root networks. It can also be produced using animal manures that otherwise have the potential to negatively impact water resources. In order to maximize the benefits of increased bioenergy use, there are several technical opportunities which must be taken advantage of as well.

Cellulosic Ethanol

Substantially increasing liquid biofuel production to meet transportation needs will require technical advancements in cellulosic ethanol, which is produced from nonedible plant parts such as leaves and stalks. Cellulosic ethanol currently requires too much heat, enzymes, and bacteria to be cost-effective. In the meantime, if conventional ethanol is to expand, new technologies or combinations of technologies will need to be put in place to make the process more efficient. Besides ethanol, there are significant opportunities to expand liquid biofuel production with advances in technologies like pyrolysis, direct catalytic conversion, and advanced gasification. Besides improving conversion technologies, advances must also occur in the area of feedstock development.

Algae

Algae are often touted as a potential feedstock for biodiesel production. As with cellulosic ethanol, it is possible to generate biofuels in this manner, but it is not yet cost-effective. In order to successfully develop algae as a biofuel feedstock, research and development need to occur. There are thousands of types of algae, for example, and some of them are better suited for biofuel production than others.

Bioenergy Challenges and the Future

The potential advantages of increased bioenergy use are numerous, but there are several hurdles to expanding bioenergy production and consumption as well. The biggest challenge facing expanded use of bioenergy is the question of sustainability. Despite being one of the largest renewable energy resources in the world, there is little good data on supply and demand of biomass in many countries. This lack of good baseline data makes it very difficult to derive sustainable bioenergy policies to guide the industry (Rosillo-Calle et al. 2007). Effective policies will need to be crafted at many levels in order to ensure the sustainability of biomass resources over the long haul. These policies will need to address soil and water health, air emissions, and the sustainability of the biomass sources themselves. Concerns about the net energy balance and greenhouse gas emissions of various types of bioenergy, especially liquid biofuels, will also need to be addressed.

Transporting and storing biomass is another very significant challenge facing the industry. Biomass has a relatively low energy density and often a high moisture content. These characteristics make moving and storing biomass a challenge—both from a technical and economical standpoint. Because of this, many observers believe that bioenergy production will happen on a distributed basis, with many small biorefineries generating energy and other bio-products across the rural landscape. This scenario presents the land use challenge of siting multiple facilities across rural communities.

Finally, bioenergy must be able to compete economically with fossil fuels and with other forms of renewable energy. If, as was the case in 2009, conventional energy prices drop, the bioenergy industry stagnates or goes into decline. Making bioenergy competitive with fossil fuels will likely require price supports, such as the monetized value of carbon reductions, in addition to higher-priced fossil fuels.

As recent volatility in ethanol prices and debates over the environmental sustainability of corn-derived ethanol have demonstrated, bioenergy is not without its challenges. Biomass is plant matter grown on a variety of soils; if biomass is not sustainably grown and harvested, the soil resources will be depleted over time. Sustainability concerns have also arisen as biomass is diverted from the food chain into the energy chain. Critics claim that this diversion contributes to increasing food prices, which negatively impacts the world's poor. These same voices contend that bioenergy development puts indirect pressure on developing countries to clear more native forests and grasslands to grow food crops, thereby accelerating deforestation and negatively impacting the environment.

Other challenges facing the bioenergy industry include the development of successful, local biomass supply chains. Developing a bioenergy industry is dependent upon the development of a biomass supply chain that can effectively and efficiently deliver an adequate supply of biomass at a reasonable price to the biorefinery. This process will require the formation of new business models and arrangements among those that grow, harvest, store, deliver, refine, and market biomass into bioenergy.

Despite the considerable challenges facing the bioenergy industry, there are a number of significant opportunities as well. New dedicated crops like miscanthus and switch grass, and short rotation woody crops like poplar and willow promise to provide environmental benefits such as carbon sequestration in addition to being excellent feedstocks. Emerging technologies like cellulosic ethanol promise to radically increase the efficiency of converting biomass to ethanol and to expand the range of potential feedstocks for bioenergy production. If these types of opportunities can be harnessed, bioenergy will continue to make a significant contribution to the expansion of renewable energy worldwide.

Andrew DANE
University of Wisconsin-Extension

See also in the *Berkshire Encyclopedia of Sustainability* Agriculture; Automobile Industry; Biotechnology Industry; Development, Rural—Developed World; Development, Rural—Developing World; Energy Efficiency; Energy Industries—Overview of Renewables; Facilities Management; Investment, CleanTech; Supply Chain Management

Further Reading

Biomass Research and Development Board. (2008). *The economics of biomass feedstocks in the United States: A review of the literature.* Retrieved October 29, 2009, from http://www.usbiomassboard.gov/pdfs/7_Feedstocks_Literature_Review.pdf

Biomass Research and Development Technical Advisory Committee & Biomass Research and Development Initiative. (2007, October). *Roadmap for bioenergy and biobased products in the United States.* Retrieved October 29, 2009, from http://www1.eere.energy.gov/biomass/pdfs/obp_roadmapv2_web.pdf

Crooks, Anthony. (2008) Ownership manual: Bioenergy study assesses four primary ownership models for biofuels. *Rural Cooperatives, 75*(1), 10–14.

Farrell, John, & Morris, David. (2008). *Rural power: Community-scaled renewable energy and rural economic development.* Retrieved October 29, 2009, from http://www.newrules.org/sites/newrules.org/files/ruralpower.pdf

The Minnesota Project. (2009, August). *Transportation biofuels in the United States: An update.* Retrieved October 29, 2009, from http://www.mnproject.org/pdf/TMP_Transportation-Biofuels-Update_Aug09.pdf

Prochnow, A.; Heiermann, M.; Plöchl, M.; Linke, B.; Idler, C.; Amon, T.; & Hobbs, P. J. (2009). Bioenergy from permanent grassland—A review: 1. Biogas. *Bioresource Technology, 100*(21), 4931–4944.

Radloff, Gary, & Turnquist, Alan. (2009). *How could small scale distributed energy benefit Wisconsin agriculture and rural communities.* Retrieved October 29, 2009, from http://www.pats.wisc.edu/pubs/97

Rosillo-Calle, Frank; de Groot, Peter; Hemstock, Sarah; & Woods, Jeremy. (2007). *The biomass assessment handbook: Bioenergy for a sustainable environment.* London: Earthscan.

Schwager, J.; Heermann, C.; & Whiting, K. (2003). Are pyrolysis and gasification viable commercial alternatives to combustion for bioenergy projects? In *Renewable Bioenergy—Technologies, Risks, and Rewards* (pp. 63–74). Bury St. Edmonds, UK: Professional Engineering Publishers.

Tilman, David; Hill, Jason; & Lehman, Clarence. (2008). Carbon negative biofuels from low input high diversity grassland biomass. *Science, 314*(5805), 1598–1600.

United States Energy Information Administration (USEIA). (2009, July). Renewable energy consumption and electricity preliminary statistics 2008. Retrieved November 6, 2009, from http://www.eia.doe.gov/cneaf/alternate/page/renew_energy_consump/rea_prereport.html

United States Environmental Protection Agency (US EPA). (2007). *Biomass conversion: Emerging technologies, feedstocks, and products.* Retrieved October 29, 2009, from http://www.epa.gov/Sustainability/pdfs/Biomass%20Conversion.pdf

Energy Industries—Geothermal

Geothermal energy is derived from the natural heat found within Earth. Its direct use as a heating source—for homes, baths, and spas—has been employed for thousands of years, and it has been utilized since the early twentieth century as a source for the generation of electricity. Its advantages are that it is clean, renewable, and abundant. The number of nations using geothermal heat is growing constantly.

The word *geothermal* literally means "Earth's heat," which is estimated to be 5,500°C at the Earth's core—about as hot as the surface of the sun. Geothermal energy can be harnessed from underground reservoirs that contain hot rocks saturated with water and/or steam. Boreholes—typically two kilometers deep or more—are drilled into the reservoirs. The hot water and steam are then piped up to a geothermal power plant, where they are used directly for heating purposes or to drive electric generators to create power for businesses and homes.

Geothermal energy is considered a renewable resource because it exploits the Earth's interior heat, which is deemed abundant, and water that is piped back to the reservoir after use and cooling. Geothermal energy can be tapped by many countries, especially those located in geologically favorable places around the world—generally volcanic areas along the major plate boundaries on the Earth's surface.

Utilization of Geothermal Energy

Geothermal energy can be utilized directly for heating purposes, food processing, fish farming, bathing, agriculture, and other applications that require heat. Geothermal energy can also be used indirectly by employing steam or heat for generating electricity. In this, geothermal is unique compared to other renewable energy technologies. It not only provides a real base-load capacity (power output that can be produced nearly continuously) for electricity generation but also presents a cleaner alternative to fossil fuels for heat production.

Direct Uses

The oldest and probably best-known application of geothermal energy—dating back to Roman times—is in baths, spas, and for heating purposes. There are many other applications for geothermal energy today: in district heating systems, including one of the world's largest in Iceland, where about 90 percent of the homes are heated with geothermal energy (Orkustofnun 2009); in thermal baths and spas; in geothermally heated fish farms; and for food processing and dehydration. Furthermore the possibility to utilize ground source (geothermal) heat pumps to either warm homes directly or through energy efficiency programs, which allows savings in energy costs, has vast possibilities for individual or industrial use.

All those "direct use" applications utilize geothermal energy produced from lower temperature water (less than 150°C) that is derived from wells 100–1,000 meters deep. As of 2005, seventy-three nations utilize geothermal energy directly, with an overall energy output of 75,900 gigawatt-hours (GWh) thermal per year (Glitnir Geothermal Energy Research 2008). (A gigawatt-hour is a measure of energy, equal to one billion watts generated for one hour; one gigawatt-hour is enough to power approximately 89 US homes for one year [calculation based on data from USEIA 2009] or 198 homes in the European Union for one year [Bertoldi et al. 2006, 12]. For more on this topic please refer to the sidebar "What's the difference between a gigawatt and a gigawatt-hour?"

on pages 4–5.) The number of nations using geothermal heat is growing constantly.

Power Generation

The utilization of geothermal energy for the generation of electricity is probably its most prominent application today. This application is a mature technology that began in a small power generation installation in Larderello, Italy, in 1904. To this day, geothermal energy is used to generate electricity and to provide heat directly in Larderello.

While the number of nations that generate electricity through geothermal energy is growing, only around twenty-four nations produced electricity from it in 2009. According to the International Geothermal Association (2009), the overall installed capacity is around 10 gigawatts (GW).

Electricity is generated from medium and high enthalpy fluids of more than 150°C derived from wells 1,000–3,000 meters deep in hot, permeable rock. The water from those wells is either used directly in a steam turbine or to heat up a secondary working fluid, which has a lower boiling-point temperature.

All applications mentioned in table 1 use water as a carrier of geothermal energy in the form of heat. This technology has been proven for hundreds of years, but it requires existing water flows in the ground. For electricity production, high temperatures are needed, but new technologies use a binary cycle system with fluids boiling at lower temperatures. This allows the generation of electricity in lower temperature regions and extends the prospects of geothermal energy utilization for electricity generation even further.

With "a higher capacity factor (a measure of the amount of real time during which a facility is used) than many other power sources" (Kagel, Bates, and Gawell 2007, i), geothermal has a huge advantage. Unlike other renewable energy resources, such as wind and solar power that are more dependent upon weather fluctuations and climate changes and have capacity factors of 20–30 percent, geothermal resources are available twenty-four hours a day, seven days a week. In real-life terms, this means that a 50 MW installation of geothermal power provides electricity for around 38,000 US households, while the same installation for wind provides electricity only for around 15,000, and solar photovoltaic only provides enough electricity for 10,000 households, based on estimates by Glitnir (2008). This gives some indication of geothermal's usefulness on an individual basis. And for any industrial user, 24–7 accessibility to electricity is needed; the ability to receive electricity only when there is enough wind or sunshine simply is not an alternative.

New technologies—including engineered geothermal / hot dry rock systems—are in various stages of development and could provide further advancements to the geothermal energy industry. They are aimed at utilizing heat from nonpermeable rocks by creating hot-water reservoirs artificially. While those systems are not commercially viable yet, they are potentially extending geothermal energy utilization dramatically, allowing electricity generation all over the world, not just in "geologically favorable areas." Further advancements to drilling technology and the economics of

Table 1: Geothermal Energy Applications

Electricity Production—Hydrothermal	Wells drilled into a geothermal reservoir produce hot water and steam from depths up to 3 kilometers The geothermal energy is converted at a power plant into electric energy, or electricity Hot water and steam are the carriers of the geothermal energy
Direct Use	Applications that use hot water from geothermal resources directly include space heating, crop and lumber drying, food preparation, aquaculture, and industrial processes Historical traces back to ancient Roman times for baths and spas
Geothermal Heat Pumps	Takes advantage of relatively constant Earth temperature as the source and sink of heat for both heating and cooling, as well as hot water provision One of the most efficient heating and cooling systems available
Hot Dry Rock Deep Geothermal / EGS*	Extracts heat by creating a subsurface fracture system to which water can be added through injection wells Water is heated by contact with the rock and returns to the surface through production wells Energy is then converted at a power plant into electric energy as in a hydrothermal geothermal system

Source: Glitnir Geothermal Energy Research (2008).
*neither commercial as of 2009

Geothermal energy—in the form of heat and carried by water—has many potential uses, both direct (heating) and indirect (power generation).

deeper drilling also provide a great hope to further growth in geothermal energy exploration and utilization.

For electricity production, further technological developments will have a huge impact on the overall advances in this sector. While the carrier medium for geothermal electricity—water—must be properly managed, the source of geothermal energy—the Earth's heat—will be available indefinitely.

Outlook for the Twenty-First Century

The overall prospects for geothermal energy utilization, either for electricity generation or direct use, are excellent. While depending heavily on political and financial support, geothermal energy represents the only real baseload capacity alternative to fossil fuels like coal and oil.

The biggest potential and prospects for the shorter term are in the direct use of geothermal energy, particularly for heating and other direct applications. With technological developments in, for example, binary systems and engineered geothermal systems, geothermal could provide electricity throughout the world.

As of late 2009, there are more than 120 projects currently under development in the United States alone, which in total could more than double that country's currently installed geothermal power capacity, with estimations of a total increase to around 10 GW by 2020. Globally, geothermal power generation capacity is expected to triple to more than 30 GW by 2020 (Geothermal Energy Association 2008; International Geothermal Association 2009; Islandsbanki 2009). While power generation will be a prominent issue in the climate change debate, the more important role may be played by geothermal energy's direct use for heating and other heat-based applications, which gives geothermal the potential of replacing the fossil fuels largely connected with pollution.

The potential for geothermal direct use is tremendous and not quantifiable, particularly with the application of new technologies, such as enhanced geothermal systems (EGS). At the same time there are local concerns about earthquakes and emissions. There is an ongoing debate about how or if work on EGS systems might possibly induce seismic activity; while the likelihood of earthquakes is limited, concerns by the general public need to be addressed by companies in the sector early in the development process. The same applies to possible emissions, although geothermal plants have minor (or no) emissions and are considered a cleaner alternative to other thermal plants.

Alexander RICHTER
Islandsbanki, Geothermal Energy Team; ThinkGeoEnergy

The author would like it noted that he wrote this article both as an employee and team member of the geothermal energy team of Islandsbanki (a bank that promotes geothermal energy) and as a writer for ThinkGeoEnergy.

See also in the *Berkshire Encyclopedia of Sustainability* Energy Industries—Overview of Renewables; Facilities Management; Investment, CleanTech; Water Use and Rights

Further Reading

Aabakken, Jorn. (Ed.). (2006, August). *Power technologies energy data book* (4th ed.). Retrieved September 14, 2009, from http://www.nrel.gov/analysis/power_databook/docs/pdf/39728_complete.pdf

Australian Geothermal Energy Association. (2009). Retrieved September 14, 2009, from http://www.agea.org.au/

Bertani, Ruggero. (2003). What is geothermal potential? *International Geothermal Association News, 53*, 1–3. Retrieved September 19, 2008, from http://iga.igg.cnr.it/documenti/IGA/potential.pdf

Bertani, Ruggero. (2007). World geothermal generation in 2007. Retrieved September 19, 2008, from http://geoheat.oit.edu/bulletin/bull28-3/art3.pdf

Bertoldi, Paolo; Atanasiu, Bogodan; European Joint Research Commission; & Institute for Environment and Sustainability. (2007). *Electricity consumption and efficiency trends in the enlarged European Union: Status Report 2006*. Retrieved January 14, 2010, from http://re.jrc.ec.europa.eu/energyefficiency/pdf/EnEff%20Report%202006.pdf

Bundesverband Geothermie E. V. (2009). Retrieved September 14, 2009, from http://www.geothermie.de/

Geothermal Energy Association. (2009). Retrieved September 14, 2009, from www.geo-energy.org

Geothermal Resources Council. (2009). Retrieved September 14, 2009, from www.geothermal.org

Glitnir Geothermal Energy Research. (2008, March). *Geothermal energy*. Retrieved October 15, 2009, from http://www.islandsbanki.is/servlet/file/FactSheet_GeothermalEnergy.pdf?ITEM_ENT_ID=5415&COLLSPEC_ENT_ID=156

Green, Bruce D., & Nix, R. Gerald. (2006). *Geothermal—The energy under our feet: Geothermal resource estimates for the United States* (NREL Technical Report 40665). Retrieved September 14, 2009, from http://www.nrel.gov/docs/fy07osti/40665.pdf

International Geothermal Association. (2009). Retrieved September 14, 2009, from http://www.geothermal-energy.org/index.php

Islandsbanki. (2009). *Financing geothermal projects in challenging times*. Retrieved September 14, 2009, from http://www.islandsbanki.is/servlet/file/Financing%20geothermal%20projects.pdf?ITEM_ENT_ID=38517&COLLSPEC_ENT_ID=156

Kagel, Alyssa. (2006). Socioeconomics and geothermal energy. Retrieved September 14, 2009, from http://www.geo-energy.org/publications/power%20points/SocioeconomicsKagel.ppt

Kagel, Alyssa; Bates, Diana; & Gawell, Karl. (2007). *A guide to geothermal energy and the environment*. Retrieved October 16,

2009, from http:// www.geo-energy.org/publications/reports/Environmental%20Guide.pdf

Lund, John W.; Freeston, Derek H.; & Boyd, Tanya I. (2005a). Direct application of geothermal energy: 2005 worldwide review. *Geothermics, 34*(6), 690–727.

Lund, John W.; Freeston, Derek H.; & Boyd, Tanya I. (2005b). Worldwide direct uses of geothermal energy 2005. In *Proceedings of the World Geothermal Congress 2005* [CD-ROM]. Reykjavik, Iceland: International Geothermal Association.

Massachusetts Institute of Technology. (2006). *The future of geothermal energy: Impact of enhanced geothermal systems (EGS) on the United States in the 21st century.* Retrieved September 30, 2008, from http://www1.eere.energy.gov/geothermal/pdfs/future_geo_energy.pdf

North Carolina Solar Center. (2008). Renewable energy portfolio standards. Retrieved September 29, 2008, from http://www.dsireusa.org/documents/SummaryMaps/RPS_map.ppt

Orkustofnun (National Energy Authority). (2009). Retrieved October 17, 2009, from http://www.os.is/page/english/

Petty, S., & Porro, G. (2007, March). *Updated US geothermal supply characterization.* Paper presented at the 32nd Workshop on Geothermal Reservoir Engineering, Stanford, CA. Retrieved on September 30, 2008, from http://www.nrel.gov/docs/fy07osti/41073.pdf

Slack, Kara, & Geothermal Energy Association. (2008). *US geothermal power production and development update: August 2008.* Retrieved September 19, 2008, from http://www.geo-energy.org/publications/reports/Geothermal_Update_August_7_2008_FINAL.pdf

Think GeoEnergy. (2009). Retrieved September 14, 2009, from www.thinkgeoenergy.com

United States Energy Information Administration (USEIA). (2009). Table 5: US average monthly bill by sector, census division, and state 2007. Retrieved January 14, 2010, from http://www.eia.doe.gov/cneaf/electricity/esr/table5.html

United States Geological Survey. (2008). Assessment of moderate- and high-temperature geothermal resources of the United States. Retrieved September 29, 2008, from http://pubs.usgs.gov/fs/2008/3082/pdf/fs2008-3082.pdf

Western Governors' Association Clean and Diversified Energy Initiative. (2006, January). *Geothermal task force report.* Retrieved September 29, 2008, from http://www.westgov.org/wga/initiatives/cdeac/Geothermal-full.pdf

World Wide Fund for Nature International. (2007). *Climate solutions: WWF's vision for 2050.* Retrieved September 14, 2009, from http://assets.panda.org/downloads/climatesolutionweb.pdf

Energy Industries—Hydroelectric

Hydropower is an important renewable source for the world's electricity supply, and even though it is a relatively mature technology, it has considerable untapped potential. Innovations include new turbine development that protects spawning fish and small hydropower plants that are more suitable for rural locales. Hydropower has social, economic, and environmental advantages and disadvantages, but the average cost of electricity production by hydropower will remain attractive.

Humans have harnessed water to perform work for thousands of years. About 2000 BCE, the Persians, Greeks, and Romans began using primitive waterwheels powered by river current for simple applications such as irrigation and grain grinding in mills; the 2,300-year-old Dujiangyan irrigation system in China's Sichuan Province is still used today. But such devices have very low efficiency and use only a small part of a stream's available energy from its velocity or motion, called the velocity head. (The vertical height that water falls is called the head; the higher the head, the greater the kinetic energy of the water as it falls on the waterwheel or turbine below. Engineers use the head and the volume of the water flow to calculate the amount of power a hydropower project can produce; this is called the potential power.)

Evolution of Hydropower

Waterwheel efficiency improved greatly with the overshot and pitchback waterwheels, which turn counterclockwise or clockwise, respectively, when water falls on the wheel. These use the weight of the water to turn the wheel and transform the water's unharnessed power into work. Turbines were the next major development in hydropower. They consist of curved blades attached to a rotating axis and are enclosed or are submerged in water. As water passes through the rotating blades, it gives up energy. Unlike waterwheels, turbines can be mounted horizontally or vertically.

The modern hydropower turbine began evolving in the mid-1700s when the French engineer Bernard Forest de Bélidor (1698–1761) wrote the four-volume *Architecture hydraulique*. In 1869 the Belgian electrician Zénobe Gramme set up the first prototype dynamo—an electric generator that produces direct current—and an electric engine. By 1881 a brush dynamo connected to a turbine in a flour mill provided street lighting at Niagara Falls, New York.

Since the 1880s, hydropower's main application has been electricity generation. By 1925, 40 percent of the world's electric energy production came from waterpower (Lejeune and Topliceanu 2002, 3), although by 2006 it had diminished to 17 percent (USEIA 2008a). This downward trend is not due to a change in global electric-energy production from waterpower, which actually increased by 74 percent during that time, but because the world's total energy production more than doubled from 1980 to 2006 (USEIA 2008a).

Importance

In 2006, the percentage of the world's electricity generated by hydropower (17 percent) represented nearly 90 percent of renewable electricity produced worldwide (Pew Center n.d.). Thus it is the most widespread form of electricity-generating renewable energy.

Since 1965, the world's total energy consumption from oil, natural gas, coal, nuclear power, and hydropower (which is the only renewable resource of those listed) increased from 44 million gigawatt-hours (GWh) to

131 million GWh (BP 2009). (A gigawatt-hour measures the total energy used over a period of time and equals 1 million kilowatt-hours; in one year, 1 GWh can power approximately 89 US homes [calculation based on data from USEIA 2009c] or 198 homes in the European Union [Bertoldi and Atanasiu 2007, 12]. For more on this topic please refer to the sidebar "What's the difference between a gigawatt and a gigawatt-hour?" on pages 4–5.) As of 2007, the world's primary energy consumption was from oil (35.6 percent), followed by coal (28.6 percent). Their consumption has been growing, but it is curbed by the growth in energy consumption from renewable sources, including hydropower (6.4 percent; BP 2009).

Hydropower consumption and production vary by country. Four countries have been both the world's largest consumers of hydropower and its largest producers. In 2008 they consumed hydropower in this order: China, 1.54 million GWh, or 18.5 percent of the world's total; Canada, 972,000 GWh, or 11.7 percent; Brazil, 957,000 GWh, or 11.5 percent; and the United States, 659,000 GWh, or 7.9 percent (BP 2009, 38). According to the United States Energy Information Administration (USEIA 2008b), those countries producing the most hydropower were China (521,000 GWh), Canada (378,000 GWh), Brazil (361,000 GWh), the United States (259,000 GWh), and Russia (153,000 GWh). Conditions such as rainfall amounts and drought can affect hydropower production; US hydropower production decreased from 2007 to 2008 because of drought-related issues. But government policies to invest in hydropower, such as China's Three Gorges project, can greatly increase production. From 2007 to 2008, China's production (20.3 percent) increased so much more than that of other countries that it skewed the world's net increase in production (BP 2009).

Although hydropower is an important source of renewable energy, it is controversial. A chief benefit is that it produces extremely small quantities of carbon dioxide (mostly from power plant construction and decaying organic matter growing in the plant's reservoir); the amount is less than those of wind, nuclear, and solar energy sources. Also, hydropower's supply is generally stable since water is abundant in many places.

One of its greatest drawbacks is cost. Hydropower's initial investment costs from dam and power plant construction are relatively high (partly because many geographic variables make project planning site specific). Other costs include the installation of (or hook up to) transmission lines, the operation and maintenance of the facility, and the financial and social costs of resettling people displaced by the dam and reservoir. Loss of agricultural land and potential damage to ecosystems are also important drawbacks (Williams and Porter 2006).

Hydropower's long-term costs, however, tend to be low because the energy source (flowing water) is renewable and free. In the United States, it costs on average 85 cents to produce 1 kilowatt-hour with hydropower, which is 50 percent cheaper than nuclear power, 40 percent cheaper than fossil fuels, and 25 percent cheaper than natural gas (Wisconsin Valley Improvement Company n.d.). A kilowatt-hour is the unit that electric companies use to charge residential customers for their energy use over time. Compilations of data for the cost of electricity generation in 2009 (USEIA 2009a, 89; 2009b) show that the average cost of electricity production by hydropower remains attractive. (See figure 1 on page 54.) Cost estimates for 2016 by the Institute for Energy Research (2010, 1) indicate that hydropower costs will remain at the same level as those for biomass and be cheaper than wind power and much cheaper than solar energy. (See figure 2 on page 55.)

Hydropower and Dams

Most twenty-first-century hydroelectric power plants consist of a reservoir for holding water, a dam that can be opened or closed to control water flow, and a power plant that generates electricity as the water flows through the turbines that spin a generator. Some power plants, called run-of-the-river plants, are built along rivers with steady or regulated flows and don't require a dam. Not all dams have power plants; only 3 percent of the estimated 75,000–79,000 US dams can generate electricity (Pew Center n.d.). The United States Department of Energy (2005) estimates that the remaining 97 percent represent up to 21,000 unused megawatts (MW) of hydropower.

Economics and safety are basic considerations in planning waterpower developments. Engineers must consider the maximum output of power at the minimum cost and construct safe and appropriate facilities that both control and manipulate the variable and uncertain natural forces of water. They must consider natural disasters and hazards due to floods and ice to ensure safety and minimize interruptions in plant operation.

Because waterpower developments frequently experience natural forces, engineers previously considered steam plants to be the most dependable prime movers. ("Prime movers" are machines that convert natural energy into work; a hydropowered steam plant transforms the energy of falling water into the steam that powers turbines.) But the interruptions in service at steam plants that were sometimes the only source of energy during fuel shortages have changed consumers' perspectives on steam plants, although they remain interested in hydropower because of high fossil fuel costs. The trend in hydropower plant design is toward simple and effective layout and greater use of stored water. (A large reservoir allows a plant to operate steadily while mitigating floods and storing water during droughts.) This

Figure 1. Average Cost of Generated Electricity by Source, 2009 (in US$/MWh)*

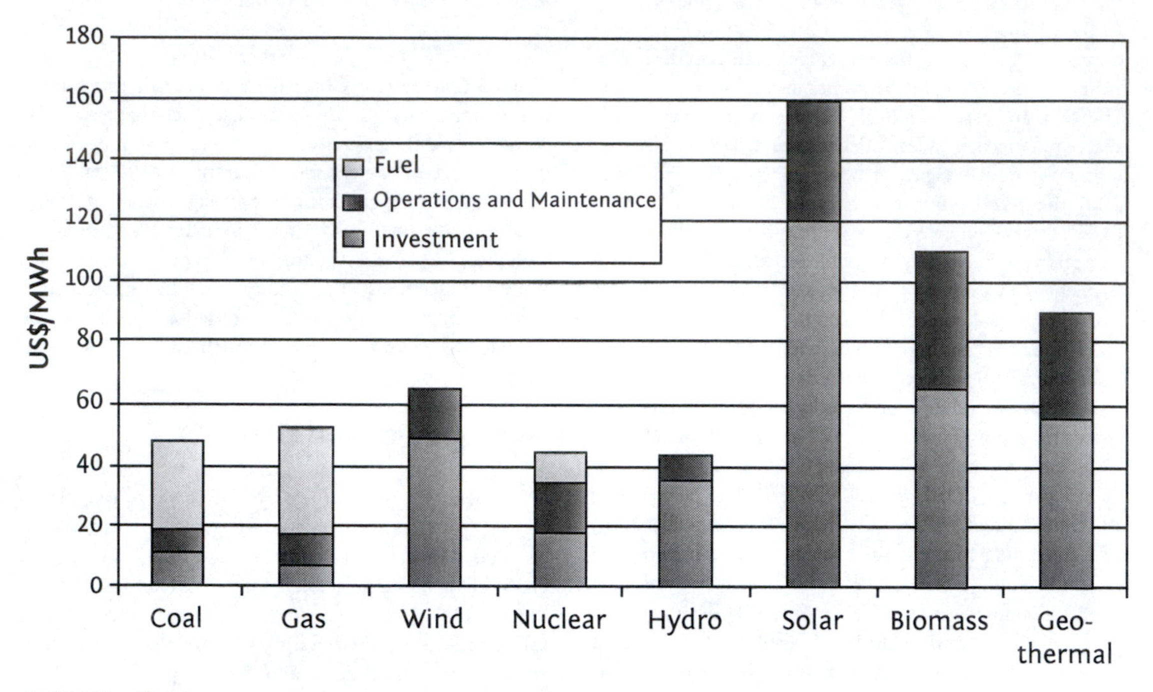

*US$/MWh = US dollars per megawatt-hour

Source: USEIA (2009a, 89; 2009b).

Hydro power is one of the least expensive energy sources for electricity generation.

trend has increased waterpower's reliability and therefore the public's appreciation of hydropower.

The likelihood of climate change and its consequences will lead to modifications of electricity generation and a demand for hydropower. But most of the "easy" potential dam sites are already developed. Because hydropower generation depends on natural conditions (availability of water and the head), every power site has unique design and construction problems, and no two waterpower developments are the same. Certain plant layouts correspond to the most-important site characteristics: head, available flow, and river topography. These characteristics are essentially interdependent and affect the dam's development, market characteristics (e.g., current technologies and regulatory policies), and type of load (i.e., whether the plant produces power constantly or on demand during peak times). These in turn affect the plant's size and the number of power-generation units. The development of future hydropower plants requires reducing costs (e.g., using low-cost roller-compacted concrete in dams) and increasing protection of the project's environment and those who live in it.

Concentrated Fall Dams

The Three Gorges power plant in China has a concentrated fall dam layout. These dams concentrate the water and provide it directly to a power station located near or in the dam. They are usually built in narrow valleys with mild slope, where concentrating the flow to a single point does not require an extraordinarily large—and extremely expensive—dam and where the spillway is located on the dam itself.

The Three Gorges dam is built at the end of Xiling Gorge in China's Hubei Province, one of three gorges on the Yangzi (Chang) River. As of 2009, it was the world's largest hydropower plant, with the dam measuring 185 meters high and 2,309 meters wide; it is expected to generate enough power to pay for its $30 billion construction cost within five to eight years. Its transmission and distribution systems will join three regional grids, making a network that will provide power from China's eastern coast to the Tibetan border in the west. The dam will also provide flood control downstream.

Figure 2. Future Cost of Generated Electricity by Source, 2016 (in US$/MWh)

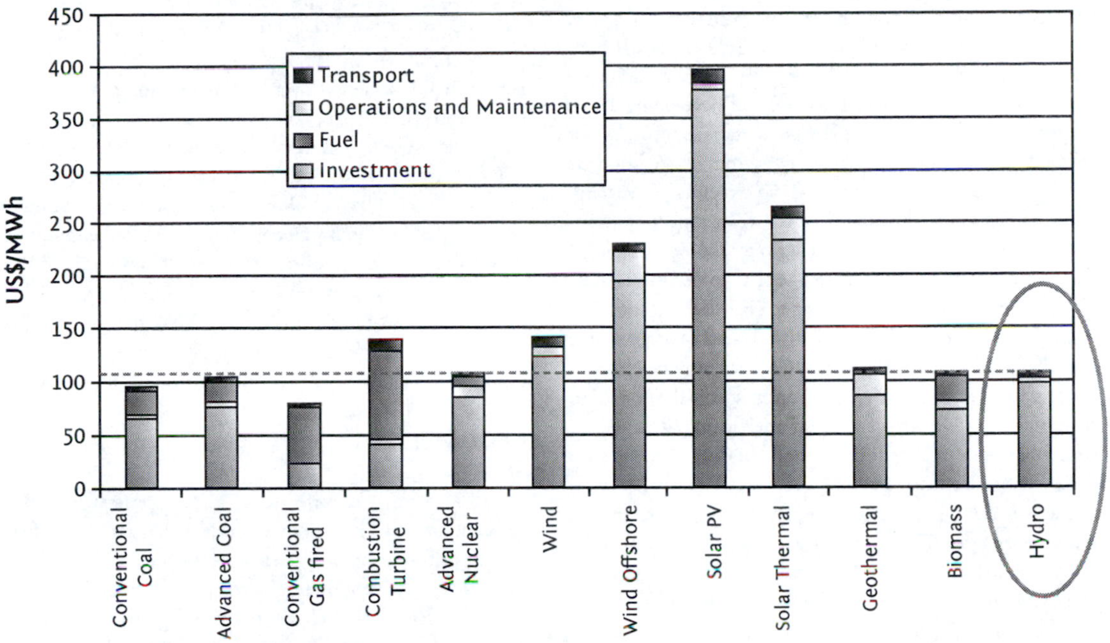

Source: Institute for Energy Research (2010, 1).

The cost of electricity generated by hydro power in 2016 is expected to remain low among renewable energy sources and competitive with all others.

Divided Fall Dams

The Grande Dixence dam in Valais Canton, Switzerland, is a divided fall dam, with the dam and the power house in separate locations. These dams are usually built in rugged terrain, where dam construction that concentrates the entire flow would be prohibitively expensive. Grande Dixence's tunnels and pumping stations collect water from glaciers, and gravity moves the stored water to four underground power stations below the dam. It is the world's tallest gravity dam, at 285 meters. Although divided fall dams harness a smaller portion of the flow than concentrated fall dams, the diverted flow can be distributed to power stations at elevations lower than the dam's base, providing a greater head. Also, divided fall dams typically generate power using a higher head and less flow, resulting in a lower cost per kilowatt generated (Lejeune and Topliceanu 2002, 8).

The environmental impact of these dam layouts has not been assessed. Their impacts depend on individual parameters such as human and animal factors, location and catchment areas, discharge value and variation, and riverbed geomorphology.

Economies of Scale

Waterpower plants can be categorized by either the head or the generation of energy. Each category is divided into three ranges: low, medium, and high head or small, medium, and large power plant, respectively. Low-head plants have low heads, no water storage, and low energy capacity. (A plant's capacity is the maximum power it can provide in a given instant.) The McArthur Generating Station on the Winnipeg River in Manitoba, Canada, is one of the largest low-head plants, capable of generating 54 MW (Manitoba Hydro n.d.). Medium-head plants have heads 30–300 meters high and large reservoirs; the Three Gorges dam is an example, with a maximum head of 113 meters (Bridle 2000). High-head plants have heads over 300 meters high, large dams, and constant electricity generation. The Grande Dixence's Fionnay power plant receives water from a net height of 874 meters and has a 294-MW capacity (Alpiq Group n.d.). The definitions of small, medium, and large power plants vary, but the size indicates the number of megawatts the plant can produce. For example, the US Department of Energy (2005) defines a small hydropower plant as producing between 100 kW and 30 MW.

Small Hydropower Plants: The Pros

One type of waterpower plant has gained prominence: the small hydropower plant (SHP). SHPs, which catch and channel water, are hydroelectric generating stations of low-power energy suitable for rural and developing areas. Engineers calculate that SHPs generate 5–8,000 kW for falls 1.5–300 meters high, for a flow from a few hundred liters per second to a few tons of cubic meters per second. Most SHPs are built on run-of-the-river power stations, with or without a small regulating tank for water storage. Engineers often need to know the river's conditions (e.g., dry season, wet season) in order to size the turbines and control production. SHPs can be connected to other power stations or used in isolated networks that provide electricity to a village, small city, or complex medical, industrial, or agricultural facility far from the principal network. Moreover, they only require a small building and one staff person without special training. Compared with other types of stations, SHPs are often completed at low cost, operate on a low budget that is usually reasonable for rural communities, and are easier to keep fully staffed, which can be an advantage in rural areas. Finally they are autonomous, starting and operating on their own without using fuel.

SHPs are often isolated networks managed by the community or a private individual or group. Energy production, however, may be uninterrupted because of the community's social events or needs. But if interruptions happen too frequently (due to water shortages, drought, malfunctions, poor maintenance, etc.), people may become frustrated and use power derived from other, more environmentally destructive fuels such as coal, or trees from deforested areas (which often occurs in the developing world). Overall, SHPs have a lower negative environmental impact than larger projects do. Their low impact and use of renewable natural resources affects the natural environment less than almost any other form of electrical energy production (International Energy Agency n.d.).

Small Hydropower Plants: The Cons

Economies of scale are important in determining a hydroelectric project's profitability. A negative aspect of SHPs is that construction costs are $1,200–$6,000 per kilowatt installed capacity (Minister of Natural Resources Canada 2004), which, despite the lower cost per project and lower operating costs, is generally three times more than large- or medium-sized hydropower plants. The risk of accidents is also higher because of the lack of continuous maintenance by qualified people.

Despite the global financial crisis that began in late 2007, new implementations or studies of hydropower plants are under way, especially small- and large-scale projects (e.g., Inga dam in the Democratic Republic of Congo [DRC], Romaine dam in Canada). Many medium-scale projects were postponed or canceled (e.g., Memve'ele in Cameroon). China continues to develop very large hydropower projects with a dam height over 300 meters and power around 6 gigawatts (GW) per plant. China's installed hydro capacity stands at about 155 GW, and it aims to increase this to 300 GW by 2020. Its total exploitable hydropower potential was estimated to be 542 GW in 2009, ranking it first in the world (Chen 2009). Brazil continues to implement projects with a dam height around 100 meters, mainly run-of-the-river plants, with power around 3 GW. Its largest hydropower plants in operation in May 2009 had an installed capacity of 50 GW; smaller-scale plants provide an additional 19 GW (Brazilian Committee on Dams 2009).

Economics

Conditions in the developed and developing world affect the economics of hydropower projects. An analysis of the cost of different power-generation systems in the United States indicates that hydroelectric plants are the most economical. These plants' mean cost of generation is only 40 percent of the cost of generation by fuel oil (Wisconsin Valley Improvement Company n.d.).

In Africa, Tanzania closed hydro plants in 2006, Kenya closed its 14-MW Masinga dam in 2009 due to recurrent droughts (Browne 2009), and the capacities of the DRC's Inga 1 and 2 dams diminished due to poor maintenance. But several large hydro projects are being planned or built there: the proposed Grand Inga complex in the DRC, which is expected to produce almost twice the electricity of the Three Gorges project (Wachter 2007); the construction of Lom Pangar dam in Cameroon; the rehabilitation and upgrading of the Kariba dam on the Zambezi River between Zambia and Zimbabwe; the construction of the Gibe III hydropower plant in Ethiopia; and the construction of the Gurara Water Transfer Project in Nigeria. In 2008, the World Bank invested more than $1 billion in small-scale and micro-hydro projects in the developing world, which displace fewer people than large projects and reduce the cost of transmitting electricity to rural areas, across vast distances, and over natural barriers such as the Sahara Desert (Browne 2009).

India's poor electricity supply stifles economic growth and reduces productivity. Hydropower in India generated nearly 37 GW in 2009 and constituted about 25 percent of India's total capacity. Each of India's twenty-eight (soon to be twenty-nine) states and seven union territories has its own power utilities that are connected through state, regional, and national transmission grids. As of 2009, two new main hydropower projects were being discussed: the 412-MW Rampur Hydropower Project and the 444-MW

Vishnugad Pipalkoti Hydropower Project. In April 2008, workers protested unsafe conditions at the Rampur construction site; nearby villagers joined them, protesting the construction of a tunnel that would threaten their only source of drinking water (Asian News International 2008). Residents and farmers protested again in September 2009, claiming the project would disturb the region's ecological balance (Asian News International 2009).

New Turbine Development

Environmental and technological considerations affect the type of equipment used in hydropower plants. For example, turbines can be hazardous to fish passing through them; their effect on fish populations can be one of the most important criteria for selecting turbines in low-head hydropower plants. Fish injuries caused by a hydroelectric turbine are often quantified by model predictions rather than actual measurements. Survival rates measured for fish passing directly through a large modern turbine are 88–94 percent. By comparison, survival rates for fish moving through a fish bypass system typically are 95–98 percent, and 95–99 percent for a spillway system. The net survival rates for fish traversing dams without bypass systems are substantially lower, because each fish has to pass through several turbines during its journey to the sea.

Fish-friendly turbine designs are vital to hydropower's future and allow fish to pass directly through the turbine without injury, eliminating the need for a separate fish bypass system. (Fish bypass systems, such as the fish ladders installed on the Ice Harbor Dam in Washington State, require additional construction, which increases the plant's initial costs.) The main manufacturers of turbines (e.g., Alstom, Alden, Voith Hydro) have identified specific mechanisms that cause injury, and they design and provide turbines to remedy these situations.

For small-scale hydropower generation, Archimedean screw turbines, which push water uphill, are a promising technology. The Archimedean screw (named for the Greek scientist Archimedes, c. third century BCE) has been used since ancient times to help irrigate crops. In the twentieth century, the German manufacturer Ritz-Atro discovered it could use the screw to generate hydroelectricity by reversing the process so that the weight of the water turns the screw. It works the same as a dynamo on a bicycle wheel. Like the cyclist turning the wheel, the water turns the screw. The screw is installed along a dam or a river weir. Water is diverted above the weir, and then flows down the screw and back into the river.

Advantages and Disadvantages

The positive social aspects of hydropower implementation include flood protection (e.g., the Three Gorges dam); the enhancement of recreational facilities (e.g., Lake Mead, between the US states of Nevada and Arizona); and job opportunities in construction and plant operation, mostly for the local populace. Sometimes the conditions created by hydropower construction enhance navigation in the affected body of water (e.g., the Three Gorges dam).

From an economic viewpoint, hydropower production saves fuel that would otherwise come from carbon-based sources. It offers a reliable and steady source of power, combined with low long-term operating and maintenance costs, and reservoirs help meet load flexibly. It also provides nations with energy independence.

Regarding environmental impacts, hydropower plants produce no waste and no atmospheric pollutants. They avoid depleting nonrenewable fuel resources and produce very few greenhouse gas emissions relative to other large-scale energy options.

Many of hydropower's positive aspects, however, can be seen as negative. The implementation of hydropower plants frequently involves resettlement, modification of local land-use patterns, and management of competing water uses and waterborne disease vectors. The same waters used for hydropower are often used in farming, fishing, navigation, flood and drought protection, and tourism (Truchon and Seelos 2004, 2). If the effects on the impacted people's livelihoods and cultural heritage are not properly addressed and managed, hydropower projects can have a detrimental social impact, particularly for vulnerable groups. Compensation of the poorest groups is inadequate in many projects (Maldonado 2009, 1–3) and nonexistent in a few, such as the Cana Brava dam in Brazil (BIC 2003, 2). Also, reservoirs of standing water increase the risks of waterborne disease, especially in tropical regions (Truchon and Seelos 2004, 3). Negative economic effects include dependence on precipitation (producing strong discharge variations); decreased reservoir storage capacity from sedimentation; and requirements for multidisciplinary involvement, long-term planning, and, often, foreign contractors and funding. Negative environmental impacts are the inundation of terrestrial habitat and the modification of aquatic habitats and hydrological regimes (e.g., the Nile River, downstream from Aswân Dam). Water quality, the temporary introduction of pollutants into the food chain, and species' activities and populations need to be monitored and managed. Hydropower plants are barriers for fish migration, especially for salmon. Poorly planned projects can contribute to global warming as much as fossil fuel sources; in its first three years, the reservoir behind the Balbina Dam in Brazil released four times the greenhouse gases of a coal plant producing the same power (Kozloff 2009).

The controversy over the Three Gorges dam highlights many issues surrounding hydropower projects around the world. The Three Gorges project had the capacity to generate 18.2 GW as of 2008 (22.5 GW by 2011), replacing the power provided by burning 36 million metric tons of coal (Panda Travel and Tour Consultant n.d.), and will prevent flooding that has killed thousands in its worst years. But the dam project has negative impacts. The reservoir behind it displaced over 1 million citizens, submerged 1,300 archaeological sites, and forever modified the ecosystem. Although the reservoir may create arable lands along its banks, lands responsible for one-tenth of China's grain production were submerged (Allin 2004). The resulting waterway provides a larger conduit for commerce with China's interior, increasing river shipping from 9 million to 45 million metric tons annually (Panda Travel and Tour Consultant n.d). The river will dump an average of 480 million metric tons of silt into the reservoir each year, but the sluice gate system for flushing the silt downstream will ensure the reservoir's value as a flood deterrent and transportation channel. Finally, the dam is designed to withstand an earthquake of 7.0 on the Richter scale (the same force that destroyed Port-au-Prince, Haiti, on 12 January 2010).

There is no system to quantifiably compare the advantages and the disadvantages of the hydropower industry. Scientific studies and multiple-criteria analyses must be performed to prepare the assessment, and those analyses should weigh the different impacts. The value of weighting will depend on who is doing the assessing, the timing, the location, and so on; who does the assessment also determines whether the focus is on the economic, social, or environmental aspects of a given hydropower project. The final assessment should combine all these aspects.

The permanent and universal criterion for the assessment of the hydropower industry's projects, implementations, operations, and related issues is their sustainability, that is, their ability to halt irreversible environmental degradation while balancing humankind's need for energy. The hydroelectricity industry cannot endanger nature's future. Even with the progress of knowledge and human behavior in the fields of biology, sociology, and economics and their related mathematical models, assessments are still unreliable, inaccurate, or missing.

André LEJEUNE
University of Liège, Belgium

See also in the *Berkshire Encyclopedia of Sustainability* Development, Sustainable; Energy Efficiency; Energy Industries—Overview of Renewables; Investment, CleanTech; Water Use and Rights

FURTHER READING

Allin, Samuel Robert Fishleigh. (2004). An examination of China's Three Gorges dam project based on the framework presented in the report of the World Commission on Dams. Retrieved January 10, 2009, from http://scholar.lib.vt.edu/theses/available/etd-12142004-125131/unrestricted/SAllin_010304.pdf

Alpiq Group. (n.d.). Grande Dixence. Retrieved January 12, 2010, from http://www.alpiq.com/what-we-offer/our-assets/hydropower/storage-power-plants/grande-dixence.jsp

Asian News International. (2008, April 17). Rampur labourers, residents demand protection of their rights. Retrieved January 14, 2010, from http://www.thaindian.com/newsportal/india-news/rampur-labourers-residents-demand-protection-of-their-rights_10035323.html

Asian News International. (2009, September 18). Farmers protest against hydro power project in Himachal. Retrieved January 14, 2009, from http://www.thefreelibrary.com/Farmers+protest+against+Hydro+Power+Project+in+Himachal.-a0208182537

Bertoldi, Paolo, & Atanasiu, Bogodan. (2007). *Electricity consumption and efficiency trends in the enlarged European Union: Status report 2006*. Retrieved January 14, 2010, from http://re.jrc.ec.europa.eu/energyefficiency/pdf/EnEff%20Report%202006.pdf

BIC. (2003). BIC factsheet: The IDB-funded Cana Brava Hydroelectric Power Project. Retrieved January 13, 2010, from http://www.bicusa.org/Legacy/Cana%20Brava%20PPA.pdf

BP. (2009). *BP statistical review of world energy: June 2009*. Retrieved October 8, 2009, from http://www.bp.com/liveassets/bp_internet/globalbp/globalbp_uk_english/reports_and_publications/statistical_energy_review_2008/STAGING/local_assets/2009_downloads/statistical_review_of_world_energy_full_report_2009.pdf

Brazilian Committee on Dams. (2009). Main Brazilian dam design III: Construction and performance. In *23rd Congress proceedings of the International Commission on Large Dams [ICOLD]*. Paris: ICOLD.

Bridle, Rodney. (2000). *China Three Gorges project*. Retrieved January 12, 2010, from http://www.britishdams.org/current_issues/3Gorges2.pdf

Browne, Pete. (2009, September 30). The rise of micro-hydro projects in Africa. Retrieved January 14, 2010, from http://greeninc.blogs.nytimes.com/2009/09/30/the-rise-of-micro-hydro-projects-in-africa/

Chen, Lei. (2009, May 11). *Developing the small hydropower actively with a focus on people's well-being, protection & improvement: Keynote speech on the 5th Hydropower for Today Forum*. Retrieved December 21, 2009, from http://www.inshp.org/THE%205th%20HYDRO%20POWER%20FOR%20TODAY%20CONFERENCE/Presentations/Speech%20by%20H.E.%20Mr.%20Chen%20Lei.pdf

Institute for Energy Research. (2010). *Levelized cost of new generating technologies*. Retrieved February 10, 2010, from http://www.

instituteforenergyresearch.org/pdf/Levelized%20Cost%20of%20New%20Electricity%20Generating%20Technologies.pdf

International Energy Agency, Small-Scale Hydro Annex. (n.d.). What is small hydro? Retrieved January 13, 2010, from http://www.small-hydro.com/index.cfm?fuseaction=welcome.whatis

Kozloff, Nikolas. (2009, November 22). Blackout in Brazil: Hydropower and our climate conundrum. Retrieved January 13, 2009, from http://www.huffingtonpost.com/nikolas-kozloff/blackout-in-brazil-hydrop_b_363651.html

Lejeune, André, & Topliceanu, I. (2002). *Energies renouvelables et cogeneration pour le development durable en Afrique: Session hydroelectricite* [Renewable energy and cogeneration for the development of sustainable development of Africa: Hydroelectricity session]. Retrieved January 13, 2010, from http://sites.uclouvain.be/term/recherche/YAOUNDE/EREC2002_session_hydro.pdf

Maldonado, Julie Koppel. (2009). Putting a price-tag on humanity: Development-forced displaced communities' fight for more than just compensation. Retrieved January 13, 2010, from http://www.nepjol.info/index.php/HN/article/view/1817/1768

Manitoba Hydro. (n.d.). *McArthur Generating Station*. Retrieved January 10, 2010, from http://www.hydro.mb.ca/corporate/facilities/gs_mcarthur.pdf

Minister of Natural Resources Canada. (2004). Small hydro project analysis. Retrieved January 14, 2010, from http://74.125.47.132/search?q=cache:u_dTQTzJtyYJ:www.retscreen.net/download.php/ang/107/1/Course_hydro.ppt+canada+minister+of+natural+resources+average+small+hydro+power+construction+cost+2004&cd=4&hl=en&ct=clnk&gl=us

Panda Travel & Tour Consultant. (n.d.). Some facts of the Three Gorges dam project. Retrieved January 12, 2009, from http://www.chinadam.com/dam/facts.htm

Pew Center on Global Climate Change. (n.d.). Hydropower. Retrieved December 18, 2009, from http://www.pewclimate.org/technology/factsheet/hydropower

Truchon, Myriam, & Seelos, Karin. (2004, September 5–9). Managing the social and environmental aspects of hydropower. Paper presented at the 19th World Energy Conference, Sydney. Retrieved January 13, 2010, from http://www.energy-network.net/resource_center/launch_documents/documents/Managing%20the%20social%20&%20environmental%20aspects%20of%20hydropower%2020.pdf

United States Department of Energy. (2005, September 8). Types of hydropower plants. Retrieved December 18, 2009, from http://www1.eere.energy.gov/windandhydro/hydro_plant_types.html

United States Energy Information Administration (USEIA). (2008a). Table 6.3: World total net electricity generation (billion kilowatt-hours), 1980–2006. Retrieved December 23, 2009, from http://www.eia.doe.gov/iea/elec.html

United States Energy Information Administration (USEIA). (2008b). World's top hydroelectricity producers, 2008 (billion kilowatt-hours). Retrieved December 23, 2009, from http://www.eia.doe.gov/emeu/cabs/Canada/images/top_hydro.gif

United States Energy Information Administration (USEIA). (2009a). *Assumptions to the annual energy outlook 2009*. Retrieved February 10, 2010, from http://www.eia.doe.gov/oiaf/aeo/assumption/pdf/0554(2009).pdf

United States Energy Information Administration (USEIA). (2009b). Figure ES4: Fuel costs for electricity generation, 1997–2008. Retrieved February 10, 2010, from http://www.eia.doe.gov/cneaf/electricity/epa/figes4.html

United States Energy Information Administration (USEIA). (2009c). Table 5: US average monthly bill by sector, census division, and state 2007. Retrieved January 14, 2010, from http://www.eia.doe.gov/cneaf/electricity/esr/table5.html

Wachter, Susan. (2007, June 19). Giant dam project aims to transform African power supplies. Retrieved January 14, 2010, from http://www.nytimes.com/2007/06/19/business/worldbusiness/19iht-rnrghydro.1.6204822.html

Williams, Arthur, & Porter, Stephen. (2006). Comparison of hydropower options for developing countries with regard to the environmental, social and economic aspects. Retrieved December 17, 2009, from http://www.udc.edu/cere/Williams_Porter.pdf

Wisconsin Valley Improvement Company. (n.d.). Facts about hydropower. Retrieved December 27, 2009, from http://new.wvic.com/index.php?option=com_content&task=view&id=7&Itemid=44

Energy Industries—Hydrogen and Fuel Cells

Since the late-twentieth century, hydrogen and fuel cells have developed as promising means of transitioning to a more sustainable economy. Hydrogen is a nonpolluting energy carrier that can be produced from almost all energy sources; fuel cells are technological devices that generate energy and heat using hydrogen or other fuels. They are clean, silent, and more efficient than combustion systems.

Over the last decade, hydrogen and fuel cells have emerged as possible resources for a more sustainable low-carbon economy. These technologies promise substantial benefits in terms of emission reductions and energy security, and thus the notion of a "hydrogen economy," defined as an energy distribution system supported by hydrogen gas, is appealing (McDowall and Eames 2006; Rifkin 2002). Several countries—the United States, the European Union, Germany, Scandinavian regions (including Iceland), and Japan—have implemented policies and research and development (R&D) plans to advance and promote these technologies, but significant technical breakthrough and cost reduction seem necessary to support actual industrial and market development (Pogutz, Russo, and Migliavacca 2009).

Hydrogen Industry

For many centuries hydrogen has fascinated scientists due to its innate properties: it is the most abundant chemical element in the universe; it is odorless, tasteless, colorless, and nontoxic; and it generates energy without exhausts when it burns. On the other hand, free hydrogen does not occur in nature; therefore it is not a primary energy source but a carrier that needs to be extracted through other energy sources.

The hydrogen industry is organized into three main stages: production, distribution and storage, and final application. Hydrogen is produced primarily from fossil fuels such as gas (48 percent), oil (30 percent), and coal (18 percent) through mature and efficient technologies such as steam reforming and coal gasification (International Energy Agency [IEA] 2007a, 1). Four percent of hydrogen production is obtained through water electrolysis, which uses electricity to split water into H_2 and O_2. Other innovative technologies—thermochemical, biological and fermentative processes, and photoelectrolysis—have been demonstrated in laboratories, but none of these solutions is economically efficient or viable yet.

With regard to the distribution phase, hydrogen can be transported in pipelines. Several thousand kilometers of hydrogen pipelines are in operation around the world, although this option is effective and convenient only for large amounts of hydrogen. Due to the low volumetric energy density of hydrogen, the investments and energy required to pump this gas are higher than for natural gas (IEA 2007a, 1). Liquid hydrogen can be transported by trucks, trains, and ships, but the amount of energy necessary for liquefaction makes these solutions more expensive than pipelines. Hydrogen distribution through either mode is more expensive than hydrocarbon logistics.

The storage of hydrogen is also complex, since it has a very low energy density by volume as a gas at ambient condition. Therefore, hydrogen must be pressurized or liquefied, which requires a large amount of energy. At present, two options are commercially feasible: gaseous compression at 350–700 bar and liquid storage at -253°C. There are also innovative solutions still under development, such as storage in solid materials. Among these, metal hydrides (a compound in which hydrogen is chemically bonded to a metal or metal-like element) are probably the most developed option, already available for niche applications, while more advanced alternatives are complex hydrides and carbon nanotubes (a cylindrical nanostructure of carbon or a carbon allotrope). All these technologies may lead to relevant breakthroughs in the future, but they still need intense R&D investments to be viable for the market and economically competitive.

Finally, with regard to market applications, large amounts of hydrogen are needed in chemical and refinery industries, but hydrogen is also used in the food industry as an additive, in food packaging, in aerospace, and in telecommunications. More than 65 million tonnes of hydrogen are produced globally every year (IEA 2007a, 1). New hydrogen markets continue to emerge, and its use in transit fleets, public transportation, decentralized power plants, and auxiliary power units has rapidly increased.

The proposal to use hydrogen in transportation in order to solve the negative effects related to hydrocarbons has been broadly discussed since the late 1990s. On the one hand, the efficiency of hydrogen as a carrier is lower when compared with other more mature technologies, and scientists and scholars have controversial opinions on the usefulness of introducing hydrogen technologies to reduce carbon dioxide (Hammerschlag and Mazza 2005). A central consideration is related to the benefit and efficiency of using renewable energies to produce carbon-free hydrogen for transportation instead of substituting coal- and oil-generated electricity.

In practice, the emissions of a hydrogen-based energy system depend on the "well to wheel" energy chain (a life cycle analysis of the efficiency of fuels used for road transportation), including primary energy sources, hydrogen production and infrastructures for transportation and storage, and end-use technologies (Simbolotti 2009). A net reduction in carbon dioxide emissions could be obtained if hydrogen is manufactured with renewable and nuclear energy, or when carbon capture and sequestration technologies are utilized at the site of hydrogen production. Finally, a sustainable transportation system needs the market diffusion of an innovative and highly efficient technology for energy conversion: fuel cells.

Fuel Cell Industry

Fuel cells are electrochemical energy conversion devices that produce electricity and heat by combining oxygen from air and a fuel, typically hydrogen or H_2-rich fuels. Fuel cells are similar to batteries, but they operate as long as fuel is supplied. The conversion process also takes place without combustion. When hydrogen is used as fuel, water and heat are the only by-products, which means that fuel cells are environmentally clean, silent, and more efficient than other combustion systems. Fuel cells' simple structure consists of two electrodes—cathode and anode—separated by an electrolyte (an electrically conductive substance). The electrolyte carries electrically charged particles from one electrode to the other. Another key element is the catalyst, such as platinum, which encourages and speeds up the reactions among the electrodes. Fuel cells can be combined into the fuel cell "stack," the core of the fuel cell system (Hall and Kerr 2003).

Fuel cells were invented in 1839 by the English scientist Sir William R. Groove, but only since the 1990s have they captured attention both from policy makers and industry as a potential answer to environmental pollution.

Fuel cells can be produced in a variety of ways, and depending on the nature of the electrolyte and the materials used, they are generally suitable for specific market applications: stationary power generation, transport, and portable devices (Lipman, Edwards, and Kammen 2004). There are primarily four different technologies used in fuel cells.

Proton exchange membrane (PEM) design uses a polymeric membrane as an electrolyte (acidic) and operates at a low temperature (80°C) with an electrical efficiency of 40–60 percent in converting fuel into power, twice that of a traditional internal combustion engine (United States Department of Energy [USDOE] 2009a). PEM fuel cells are used mostly to power vehicles and produce residential energy.

Molten carbonate fuel cells (MCFCs) use immobilized liquid molten carbonate (alkaline). The operating temperature is high (650°C) and the efficiency is around 46 percent, but if waste heat is recycled, the rate can rise to 80 percent (USDOE 2009a). Due to their characteristics, MCFCs are typically used for power stations.

Solid oxide fuel cells (SOFCs) use solid-state ceramic electrolyte (alkaline) and operate at a very high temperature (1,000°C) with an electrical efficiency of around 35–43 percent (USDOE 2009a). This technology is utilized for auxiliary power units, electric utility, and large distributed generation.

A last example of popular fuel cell technology is the direct methanol fuel cell (DMFC). This is a relatively new technology that uses a polymer membrane as an electrolyte. The operating temperature is between 50°C and 120°C, and the efficiency is around 40 percent (Fuel Cell Markets 2010). DMFCs can use pure methanol as fuel and are typically suitable for portable applications (for example, cell phones and laptops).

Several reasons exist to consider fuel cells as a disruptive technology with strong potentialities (Nygaard and Russo 2008). First, they produce power with a higher efficiency rate and almost zero environmental impact when compared to other power systems used in internal combustion engines. Second, their flexibility and modularity can efficiently generate power in a variety of system sizes.

On the other hand, fuel cells are emerging technology that must cope with different technical obstacles in order to achieve market acceptability. At the same time, they are in competition with mature technologies, like the internal combustion engine or turbines, and other alternative devices, such as electric batteries. Two specific barriers diminish the effectiveness of fuel cell performance: their reliability and cost. Reliability of fuel cells is influenced by several contingencies, including the start-up temperature, fuel purity, durability of materials and components, and degree of humidification. In any event, the technology requires significant R&D investments to become competitive with other viable solutions.

To improve fuel cell functionality in transportation, the costs per kilowatt (kW) must be dramatically reduced (a competitive cost is around $60–$100 per kW, while the current cost of a PEM stack exceeds $1,000 per kW). (For more on this topic please refer to the sidebar "What's the difference between a gigawatt and a gigawatt-hour?" on page 4–5.) In the realm of stationary applications, however, the cost of fuel cells is expected to become competitive within a few years; while the current cost of the fuel cell stack is about $5,000 per kW, the target is an installation cost of $1,500 per kW (IEA 2007b; Simbolotti 2009). In order to reach these objectives, major reductions can be obtained abating manufacturing costs through economy of scale and learning curves (from prototype to continuous production).

As of 2009, a few hundred fuel cell–powered cars and buses operate worldwide in the transport industry; the United States has two hundred fuel cell vehicles and twenty hydrogen buses. In 2008 the production of small power-generation fuel cells was about 4,000 units, 95 percent of which were PEM (Adamson 2009). The amount of large power systems installed has reached 20 megawatts (MW), and the average size of these technologies has increased to one MW per unit (Adamson 2008). Moreover, the number of niche transport markets, such as marine applications and auxiliary power units, is growing. In particular, sales of warehouse vehicles and forklift applications have stimulated early market growth, including the distribution of a large number of demonstration units in North America and Europe.

Outlook for the Future

To achieve the goal of an environmentally sustainable, secure, and competitive energy system, there is the need for a rapid shift to new technological paradigm. Hydrogen as a carrier and fuel cells as conversion technologies have a unique position within a clean energy and transport system. According to HyWays (2008), an integrated project to develop hydrogen energy in the European Union, oil consumption could be reduced 40 percent by 2050 if 80 percent of cars were run on hydrogen. The transition, however, will take a long time. Path dependency on dominant market solutions and technological obstacles require huge public and private investments to be outbalanced. According to the US roadmap implemented by the US Department of Energy (2002), only a coordinated agenda will allow a real change toward hydrogen.

Several countries and governments have developed policies and roadmaps to implement hydrogen and fuel cells technologies. The European Union hydrogen strategy was launched in 2003 with the report *Hydrogen Energy and Fuel Cells—A Vision of Our Future* (European Commission 2003). In 2004 a formal public and private network called European Hydrogen and Fuel Cells Technology Platform (HFP) was launched with the goal to prepare and direct an effective strategy for implementing a hydrogen-oriented energy economy (HFP 2005). In 2007, the HFP endorsed a development strategy and the implementation plan (HFP 2007), defining specific scenarios and setting targets for 2020 for hydrogen and fuel cells technologies, both for transport and stationary energy systems. The US Department of Energy has launched a broad hydrogen program; these technologies are considered an important part of the department's energy technology portfolio, addressing critical energy challenges like reducing carbon dioxide emissions and ending dependence on imported oil (USDOE 2006; USDOE 2009b). This program includes R&D support activities, demonstration projects and technology validation, codes and standards definition, and international cooperation. Other countries such as the United Kingdom, Canada, Germany, Scandinavian regions (including Iceland), Japan, and New Zealand have identified these technologies as desirable options to create a low-carbon, sustainable economy, and are developing similar roadmaps, policies, and plans.

But both the hydrogen and fuel cells industries look lively and dynamic: over two hundred hydrogen refueling stations are now operating around the world (sixty in the United States), compared with only fifty in 2003; many demonstration projects have been launched by central and local governments in partnership with industrial associations and firms. From 1996 to 2008, more than ten thousand fuel cells patents were issued, and a steady growth in strategic alliances was registered (Pogutz, Russo, and Migliavacca 2009).

At present, there are four major scientific and technical challenges that must be tackled for a transition to a hydrogen-based energy system (Blanchette 2008; Edwards et al. 2008): lowering the cost of hydrogen production to a level comparable with oil; development of carbon dioxide–free hydrogen production technologies from renewable energies at competitive costs; development of infrastructures for distribution and viable storage systems for vehicles and stationary applications; and dramatic improvement in the fuel cells efficiency and costs. Whether the "hydrogen economy" will happen is not certain, but surely the hydrogen and fuel cell industries will contribute to the revolution for a more sustainable energy system.

Stefano POGUTZ and Paolo MIGLIAVACCA
Bocconi University

Angeloantonio RUSSO
Parthenope University

See also in the *Berkshire Encyclopedia of Sustainability* Automobile Industry; Energy Efficiency; Energy Industries—Overview of Renewables; Investment, CleanTech; Public Transportation; Smart Growth; Telecommunications Industry

Further Reading

Adamson, Kerry-Ann. (2008). 2008 large stationary survey. Retrieved December 14, 2009, from http://www.fuelcelltoday.com/media/pdf/surveys/2008-LS-Free.pdf

Adamson, Kerry-Ann. (2009). Small stationary survey 2009. Retrieved December 14, 2009, from http://www.fuelcelltoday.com/media/pdf/surveys/2009-Small-Stationary-Free-Report-2.pdf

Barbir, Frano. (2005). *PEM fuel cells: Theory and practice.* Burlington, MA: Elsevier Academic Press.

Blanchette, Stephen, Jr. (2008). A hydrogen economy and its impact on the world as we know it. *Energy Policy, 36*(2), 522–530.

Edwards, P. P.; Kutznetsov, V. L.; David, W. I. F.; & Brandon, N. P. (2008). Hydrogen and fuel cells: Towards a sustainable energy future. *Energy Policy, 36*(12), 4356–4362.

European Commission. (2003). *Hydrogen energy and fuel cells: A vision of the future. EUR 20719.* Retrieved December 14, 2009, from http://ec.europa.eu/research/energy/pdf/hlg_vision_report_en.pdf

European Hydrogen and Fuel Cell Technology Platform (HFP). (2005). *Deployment strategy.* http://ec.europa.eu/research/fch/pdf/hfp_ds_report_aug2005.pdf#view=fit&pagemode=none

European Hydrogen and Fuel Cell Technology Platform (HFP). (2007). *Implementation Plan – Status 2006.* http://ec.europa.eu/research/fch/pdf/hfp_ip06_final_20apr2007.pdf#view=fit&pagemode=none

Fuel Cell Markets. (2010). DMFC: Direct methanol fuel cell portal page. Retrieved February 25, 2010, from http://www.fuelcellmarkets.com/fuel_cell_markets/direct_methanol_fuel_cells_dmfc/4,1,1,2504.html?FCMHome

Hall, J., & Kerr, R. (2003). Innovation dynamics and environmental technologies: The emergence of fuel cell technology. *Journal of Cleaner Production, 11*(4), 459–471.

Hammerschlag, Roel, & Mazza, Patrick. (2005). Questioning hydrogen. *Energy Policy, 33*(16), 2039–2043.

Huleatt-James, Nicholas. (2008). 2008 hydrogen infrastructure survey. Retrieved August 4, 2008, from http://www.fuelcelltoday.com/media/pdf/surveys/2008-Infrastructure-Free.pdf

HyWays. (2008). *European hydrogen energy roadmap. Action plan: Policy measures for the introduction of hydrogen energy in Europe.* Retrieved November 12, 2009, from http://www.hyways.de/docs/Brochures_and_Flyers/HyWays_Action_Plan_FINAL_FEB2008.pdf

International Energy Agency (IEA). (2007a). *IEA energy technology essentials: Hydrogen production & distribution.* Retrieved February 5, 2010, from http://www.iea.org/techno/essentials5.pdf

International Energy Agency (IEA). (2007b). *IEA energy technology essentials: Fuel cells.* Retrieved February 25, 2010, from http://www.iea.org/techno/essentials6.pdf

Lipman, Timothy E.; Edwards, Jennifer L.; & Kammen, Daniel M. (2004). Fuel cell system economics: Comparing the costs of generating power with stationary and motor vehicle PEM fuel cell systems. *Energy Policy, 32*(1), 101–125.

McDowall, William, & Eames, Malcolm. (2006). Forecasts, scenarios, visions, backcasts and roadmaps to the hydrogen economy: A review of the hydrogen futures literature. *Energy Policy, 34*(11), 1236–1250.

Nygaard, Stian, & Russo, Angeloantonio. (2008). Trust, coordination and knowledge flows in R&D projects: The case of fuel cell technologies. *Business Ethics: A European Review, 17*(1), 24–34.

Pogutz, Stefano; Russo, Angeloantonio; & Migliavacca, Paolo O. (Eds.). (2009). *Innovation, markets and sustainable energy: The challenge of hydrogen and fuel cells.* Cheltenham, UK: Edward Elgar.

Rifkin, Jeremy. (2002). *The hydrogen economy: The creation of the world-wide energy web and the redistribution of power on Earth.* New York: Putnam.

Simbolotti, Giorgio. (2009). The role of hydrogen in our energy future. In Stefano Pogutz, Angeloantonio Russo, & Paolo Ottone Migliavacca (Eds.), *Innovation, markets and sustainable energy: The challenge of hydrogen and fuel cells* (pp. 3–19). Cheltenham, UK: Edward Elgar.

Solomon, Barry D., & Banerjee, Abhijit. (2006). A global survey of hydrogen energy research, development and policy. *Energy Policy, 34*(7), 781–792.

United States Department of Energy (USDOE). (2002). *National hydrogen energy roadmap. Toward a more secure and cleaner energy future for America.* Retrieved December 12, 2009, from http://www.hydrogen.energy.gov/pdfs/national_h2_roadmap.pdf

United States Department of Energy (USDOE) and United States Department of Transportation. (2006). *Hydrogen posture plan: An integrated research, development and demonstration plan.* Retrieved May 10 2008, from http://hydrogen.energy.gov/pdfs/hydrogen_posture_plan_dec06.pdf

United States Department of Energy (USDOE). (2009a). Comparison of fuel cell technologies. Retrieved January 26, 2010, from http://www1.eere.energy.gov/hydrogenandfuelcells/fuelcells/pdfs/fc_comparison_chart.pdf

United States Department of Energy (USDOE). (2009b). Multi-year research, development and demonstration plan: Planned program activities for 2005–2015. Retrieved December 14, 2009, from http://www1.eere.energy.gov/hydrogenandfuelcells/mypp/

Energy Industries—Natural Gas

Natural gas, a nonrenewable but relatively clean fossil fuel, accounts for almost 25 percent of worldwide energy consumption. Recent technological advances have made gas from unconventional sources more economical. With appropriate legislation, the substitution of natural gas for coal in the power generation industry can contribute to a more sustainable energy economy.

Natural gas is a fossil fuel, and thus its supply is finite. The use of natural gas to produce, for example, electricity also gives rise to carbon emissions that contribute to global warming. Moreover, the uncontrolled release of natural gas into the atmosphere contributes to global warming since methane, the principal component of natural gas, is itself a greenhouse gas. Yet because natural gas is the cleanest of the fossil fuels and is more abundant than previously believed, natural gas can contribute to a more sustainable energy economy.

Environmental Impacts

Environmental consequences are associated with the exploitation and use of natural gas resources. Natural gas that is inadvertently vented from wells, pipelines, and processing facilities directly contributes to global warming since methane has a global warming potential that is approximately twenty-one times that of carbon dioxide (United States Energy Information Administration [USEIA] 2008b). Adverse consequences to water supplies and wildlife habitats can also occur from the exploitation of natural gas resources.

The emissions associated with using natural gas as a fuel, although lower than those stemming from burning coal, are still nonzero. On average a traditional steam turbine using natural gas as a fuel releases 0.54 kilograms of carbon dioxide emissions for each kilowatt-hour (kWh) of electricity produced; this is approximately 43 percent lower than when a steam turbine produces electricity by using coal as a fuel (USEIA 2010a, 54, 106). When electricity is produced using a modern natural-gas combined-cycle generating unit (under this technology the waste heat arising from the turbine used to produce electricity is captured and used to create steam and generate even more electricity), the carbon emissions per kWh are about 57 percent lower than when coal is used as a fuel. (See table 1.) (For a reference to the various measurements used in this article, please refer to the sidebar "What's the difference between a gigawatt and a gigawatt-hour?" on pages 4–5.)

It is technically feasible to sequester a substantial proportion of the carbon emitted when natural gas is used as a fuel. According to the USEIA (2009b, 108), the capital costs of a power plant that captures and sequesters its carbon emissions are about 50 percent lower when the plant is fueled by natural gas as compared with coal.

Consumption

In 2006, the latest year for which data are available, natural gas accounted for about 23 percent of worldwide energy consumption (USEIA 2009a, 307). People increasingly appreciate the environmental benefits of natural gas compared with coal in the electric power sector, making natural gas the "go-to" resource for new electric generation facilities. In 2009, over 50 percent, or 12.3 gigawatts (GW), of all new generation plants planned in the United States were gas-fired units (USEIA 2010b). The USEIA believes that the costs per unit of capacity for these power plants are between one-third and one-half the costs of new nuclear plants or coal-powered plants (2009b, 108). Moreover, in contrast to coal, the owners of these plants will be well

TABLE 1. **Carbon Dioxide (CO_2) Emission Factors**

Fuel	CO_2 per Gigajoule (in Kg)	CO_2 per kWh (in Kg)*	Possible Conversion Technology	Typical Power Conversion Efficiency Factors (%)	CO_2 per KWh for Fuel and Power Conversion Technology (in Kg)
Bituminous coal	97.65	0.31	Steam	34	0.94
Distillate fuel oil	76.76	0.23	Combined cycle	31	0.81
Residual fuel oil	82.72	0.27	Steam	33	0.82
Natural gas	55.68	0.18	Combined cycle	45	0.40

*assuming 100 percent efficiency in power conversion.

Note: The figures in the table were converted to metric units.

Source: Based on data reported by USEIA (2010a, 54, 106).

The amount of carbon dioxide emitted by fossil fuels used for power generation varies by the type of fuel and power conversion technology. The higher the efficiency of the technology, the less carbon dioxide produced per kilowatt-hour produced (far right column). A steam generator using coal, for example, emits more than twice as much carbon dioxide per kilowatt-hour as a combined cycle generator using natural gas.

situated if cap-and-trade legislation, which limits emissions and allows companies to trade emissions credits, were passed and carbon emissions were penalized.

In the United States, this increase in power generation fueled by natural gas sharply contrasts with the situation in the 1970s when the Powerplant and Industrial Fuel Use Act (FUA) was enacted by the federal government in response to concerns over the adequacy of the natural gas supply. The FUA restricted the construction of power plants using natural gas and the industrial use of natural gas in large boilers. But in 2010, most scholars accept that the natural gas supply issues of the 1970s were largely due to price regulations that discouraged natural gas drilling and production. The FUA was repealed in 1987; as a result, between 1988 and 2002, natural gas consumption by the power and industrial sectors in the United States increased by approximately 45 percent (USEIA 2008a).

A major impediment to the additional use of natural gas in the power sector is that it typically has a premium price relative to coal. For example, the USEIA reports that the average cost of coal to electric utilities in the United States was US$1.94 per gigajoule (GJ) in 2008, while the average cost of natural gas was US$8.74 per GJ (USEIA 2010a, 39; USEIA 2010c). While the technology to produce electricity favors gas over coal (the technology is more efficient and also allows electricity producers to change output more quickly when the demand for electricity changes), the difference in the cost of the fuels is a major hurdle to the increased penetration of natural gas in the power sector.

There is without doubt some who would argue against the use of any fossil fuel to produce electricity; some would argue that 100 percent of electricity should come from wind and solar power. The reality is that the stability of a power system requires that demand equals supply at all times. Unlike solar- and wind-powered generating units, those fueled by natural gas have the flexibility to help maintain this balance.

How Abundant Is Natural Gas?

Natural gas is a nonrenewable resource. In its latest worldwide assessment of energy resources, the United States Geological Survey (USGS 2000) estimates that the world's remaining undiscovered conventional natural gas resources are approximately 147 trillion cubic meters (5,200 trillion cubic feet). Annual worldwide consumption of natural gas is about 3 trillion cubic meters. One widely accepted view is that the level of natural gas resources is a fixed amount that will be depleted over time, even as new resources are discovered and drilled (Dahl 2004, 336–337). For example, the researcher R. W. Bentley indicates that conventional gas production will peak by about 2030 and then decline rapidly due to depletion. According to Bentley, as production declines, "'demand' will not be met; users may have to ration; prices will rise; there is likely to be inflation, recession, and international tension" (Bentley 2002, 203).

An Alternative View

The fixed resource view of natural gas supplies has been called into question in the twenty-first century. Researchers increasingly recognize that viewing the resource base as fixed overlooks its heterogeneity in terms of cost. Under this alternative view, the resource base is thought of as a pyramid with a relatively small amount of high-quality resource that is inexpensive to find, develop, and extract at its top. Below are increasing amounts of lower quality resource that are much more costly to develop given existing technology. (See figure 1.) The resource base has three measures: the economically recoverable component, the technically recoverable component, and the "in-place" measure. The in-place measure of the resource base is the total quantity of the resource. The technically recoverable component is the quantity that can be exploited using current technology. The economically recoverable component is that portion of the technically recoverable component that can be exploited profitably under current market conditions. Resources that cannot be profitably exploited are known as subeconomic.

One important implication of the pyramid view of the resource base is that future supplies of oil and gas critically depend on the introduction of new technologies that can "move" the subeconomic portion of the pyramid into the economic component.

The heterogeneity in the natural gas resource base vividly illustrates the validity of the pyramid concept. Since the inception of the natural gas industry, its mainstay has been natural gas trapped in domelike rock formations. This gas is known as conventional gas. Natural gas from these reservoirs can be produced economically without large-scale stimulation to liberate the gas or sophisticated production equipment. Natural gas from conventional deposits deeper than 4,500 meters was once considered unconventional, but that label has faded as the average depth of wells has increased.

Unconventional Gas Sources

Other sources of natural gas—such as coal-bed methane (gas trapped within coal seams), tight gas (gas in very impermeable reservoirs with very low porosity), and gas shales (gas found in the shale rock created during the Devonian period of the Paleozoic era)—are referred to as unconventional gases. Compared with a representative conventional gas well in the United States from the 1960s, for example, the gas from these sources tends to have low ultimate recovery per well and low average production rates. Production depends on the application of horizontal

Figure 1. Pyramid View of Energy Resources

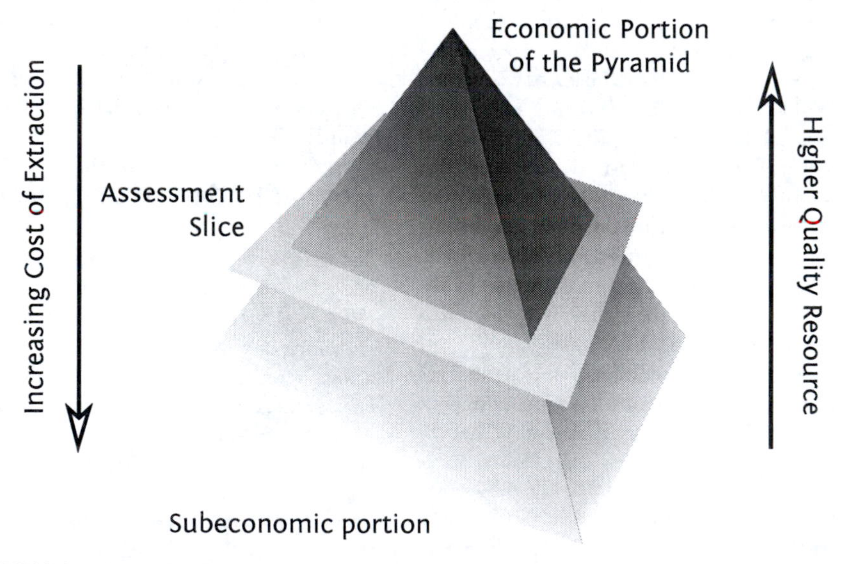

Source: Courtesy of USGS, from Ahlbrandt and McCabe (2002).

The resource base pyramid. The technically recoverable component of the pyramid (the middle) is that which can be exploited using current technology; the economically recoverable component (the top) is that portion of the technically recoverable component that can be exploited profitably under current market conditions.

drilling technology and artificially fracturing the methane-bearing rock. With a horizontal well, the operator drills vertically for up to thousands of meters and then steers the drill bit until it is sideways with the methane-bearing layer of rock. In terms of the environment, horizontal drilling makes it possible to exploit the energy resources that may lie beneath an environmentally sensitive location by drilling from a location that may be a kilometer or more away. With respect to fracturing technology, there are environmental risks if the chemicals used to help fracture the shale enter the water table. (This technology, commonly known as "fracking," is discussed in the article "Shale Gas Extraction.") One well may use 2 to 4 millions of gallons of water that often ends up containing salt, hydrocarbons, and fracturing fluids (Groat 2009, 20). While these adverse impacts can possibly be mitigated, under current US policy, hydraulic fracturing is exempt from Safe Drinking Water Act provisions (Groat 2009, 21).

A large proportion of the world's remaining natural gas resources are accounted for as unconventional gas. According to the USGS and the Minerals Management Service of the Department of the Interior, the technically recoverable natural gas resources of the United States were approximately 49 trillion cubic meters as of 1 January 2007 (USEIA 2009b, 115). To put this number in perspective, this quantity is about eighty-five times the level of domestic production around that time. Approximately 18 trillion cubic meters of these resources are from unconventional sources such as gas from tight formations, gas shale, and coal-bed methane. These resources have historically been viewed as costly relative to conventional gas and thus are sometimes considered midlevel on the resource pyramid. Resources at the base of the pyramid include gas hydrates, which are molecules of methane trapped within molecular structures, or "cages," of ice. A clump of these hydrates looks much like a snowball—a "snowball" that if set on fire burns with a clean blue flame.

In contrast to conventional oil and gas, the world's gas hydrate resources are geographically dispersed. (See figure 2.) They are also very large. Estimates of the world's *in-place* natural gas hydrate resources range widely from about 2,830 trillion to 7,640,000 trillion cubic meters (Collett 2009, 1, 2). These values are between 1,000 and 6,400 times the world's current level of natural gas production. And while large, they do not address the issue of recoverability, that is, how much of the gas we can expect to actually produce. There are potential hazards associated with the production of natural gas from hydrates (USEIA 1998, chapter 3). These risks include the inadvertent release of large amounts of methane. Among the interesting developments on this issue are the reports of approximately 2.4 trillion cubic meters of undiscovered, technically recoverable gas from natural gas hydrates on Alaska's North Slope

Figure 2. Likely Locations of Gas Hydrates Worldwide

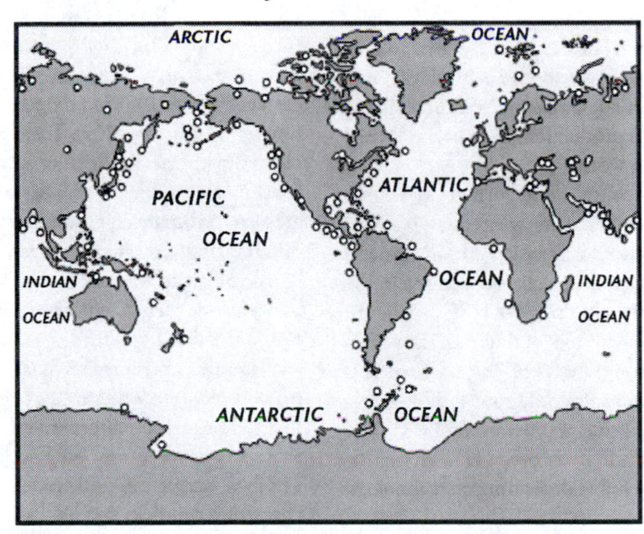

Source: United States Geological Survey. (n.d.). Gas hydrates. Retrieved February 2, 2010, from http://energy.usgs.gov/other/gashydrates/

Gas hydrates, icelike solids formed from a combination of water and natural gas, are extremely abundant around the world. Economically feasible extraction, however, remains elusive, at least in the immediate future.

(USGS 2008, 3). In 2009 the USGS reported that the US Gulf of Mexico contains gas hydrates that can produce gas using existing technology; this may be an important finding given that this area already has the necessary infrastructure (e.g., pipelines and drilling platforms) to produce the gas. Despite these developments, 2020 is probably the earliest that the production of gas from hydrates is commercialized in significant quantities.

The Shale Gas Revolution

Shale gas is largely comprised of methane trapped in shale rock. The gas is liberated from the rock and will flow through the wellbore, or drilled hole, to the surface when the rock is fractured. It is distinct from unconventional sources of oil such oil shale and oil sands, both of which are very carbon intensive in terms of extraction since both require heat to extract the resource. At one time, shale gas was considered a marginal supply source because the low permeability of the shale rock contributes to production levels that are largely subeconomic. Indicative of this is the fact the USEIA did not report shale gas reserves until 2007; since 1989 its focus was the reserves of conventional gas and coal-bed methane. This perception began changing in the late 1990s when natural gas producers in Texas reported success in developing the Barnett shale field by using horizontal drilling and hydraulic fracturing technology. By drilling wells horizontally, the rate of extraction is higher since the rock formations are also horizontally oriented.

Hydraulic fracturing technology uses high-pressure water or chemicals to induce fractures in the rock. These fractures liberate the gas from the rock and can lead to production levels that justify this high-cost well drilling. Refinement of these technologies has significantly altered the economics of shale gas production, both in the United States and throughout the world.

As of 2010, the Marcellus shale formation (which runs from New York through the western portion of Pennsylvania, into the eastern half of Ohio, and through West Virginia) is the epicenter of shale gas activity in the United States. As recently as 2002, the USGS believed that the undiscovered gas resources of the Marcellus shale are between 22.6 billion and 101.9 billion cubic meters of gas, with a mean of about 56.6 billion (USGS 2003, 1). By 2008, some believed that approximately 1.4 trillion cubic meters of shale gas could be recoverable, with significantly higher quantities possible depending on technology (Messer and Fong 2008).

Unlike conventional gas, there is little or no exploration risk (the financial risk of drilling a "dry hole") associated with shale gas—the shale deposits are known, and the only issue is locating the "sweet spots." Moreover, in contrast to gas located on the North Slope of Alaska (a "conventional but remote gas" not yet economically feasible to produce because it lacks a pipeline infrastructure), a large portion of the shale gas resources are located near market centers in the northeastern United States.

Worldwide Implications

The USEIA believes that advances in unconventional gas technology have worldwide implications for the natural gas supply. For example, both Europe and Asia are believed to have large quantities of gas shale resources that could be exploited (Sweetnam and USEIA 2009). According to Nobuo Tanaka, the executive director of the International Energy Agency (2009), "Unconventional gas is unquestionably a game-changer in North America with potentially significant implications for the rest of the world." Tony Hayward, the chief executive of BP, one of the world's largest energy producers, has indicated that BP is applying the gas shale technology to other regions, including North Africa, the Middle East, Europe, China, and Latin America; he has further stated that unconventional gas resources could contribute an extra 4,000 trillion cubic feet to reserves over the next few years (Watson 2009). BP is not alone in its assessment: BG, Petronas, StatoilHydro, Shell, and ConocoPhillips have expanded their activity in unconventional gas plays.

Exxon is also active in applying the technology and has obtained leases in prospective shale-gas formations in Germany, Hungary, and Poland (Gold 2009). If these prospects are developed, Europe's dependence on gas supplies from Russia could be significantly reduced. It is therefore not surprising that Alexander Medvedev, deputy chief executive of Gazprom (the world's largest natural gas extractor), has dismissed the significance of unconventional resources (Watson 2009). There is some truth to his skepticism; constraints on drilling and the absence of attractive fiscal terms are likely to be barriers to development. There are also legitimate environmental concerns about the effect of fracturing on water supplies.

The effect of the increase in natural gas supply on carbon emissions has recently been analyzed by the think tank Resources for the Future. It concludes that the additional gas supplies and improvements in drilling technologies make it possible for natural gas to be an important bridge fuel (a low-carbon alternative to coal used while renewable energy sources are being developed)—but only if either a carbon tax or cap-and-trade program are implemented (Brown, Krupnick, and Walls 2009).

Natural gas is a nonrenewable fossil fuel, and its use contributes to global warming. Yet because it is the cleanest fossil fuel, has lower sequestration costs relative to coal, and is far more abundant than previously believed, the substitution of natural gas for coal in the electric power sector can make a major contribution to a more sustainable world. This transition has already begun; it will accelerate only if carbon emissions into the atmosphere are priced into all fuels, that is, when either a carbon tax is imposed or cap-and-trade legislation is passed.

Kevin F. FORBES
The Catholic University of America

Adrian DiCianno NEWALL
Energy Consultant

See also in the *Berkshire Encyclopedia of Sustainability* Cap-and-Trade Legislation; Energy Efficiency; Energy Industries—Coal; Energy Industries—Overview of Renewables; Mining; True Cost Economics

Further Reading

Ahlbrandt, Thomas S., & McCabe, Peter J. (2002, November). Global petroleum resources: A view to the future. Retrieved February 2, 2010, from http://www.agiweb.org/geotimes//nov02/feature_oil.html

Bentley, R. W. (2002). Global oil & gas depletion: An overview. Retrieved February 1, 2010, from http://www.peakoil.net/publications/global-oil-gas-depletion-an-overview

Brown, Stephen P. A.; Krupnick, Alan J.; & Walls, Margaret A. (2009, December). *Natural gas: A bridge to a low-carbon future?* (Issue Brief 09–11). Retrieved February 1, 2010, from http://www.rff.org/RFF/Documents/RFF-IB-09-11.pdf

Collett, Timothy S. (2009, July 30). *Statement of Dr. Timothy S. Collett, research geologist, US Geological Survey, US Department of the Interior, before the House Committee on Resources, Subcommittee on Energy and Mineral Resources, on unconventional fuels II: The promise of methane hydrates* [Transcript]. Retrieved February 1, 2010, from http://resourcescommittee.house.gov/images/Documents/20090730/testimony_collett.pdf

Dahl, Carol A. (2004). *International energy markets: Understanding pricing, policies, and profits.* Tulsa, OK: PennWell Books.

Gold, Russell. (2009, July 13). Exxon shale-gas find looks big. Retrieved February 3, 2010, from http://online.wsj.com/article/SB124716768350519225.html

Groat, Chip. (2009). Groundwater and unconventional natural gas development: Lessons from the Barnett and Haynesville shales. Retrieved February 16, 2010, from http://www.epa.gov/region6/water/swp/groundwater/2009-gws-presentations/06-gw-and-unconventional-natural-gas-development_groat.pdf

International Energy Agency. (2009). Press release (09)16: The time has come to make the hard choices needed to combat climate change and enhance global energy security, says the latest IEA World Energy Outlook. Retrieved March 5, 2010, from http://www.iea.org/press/pressdetail.asp?PRESS_REL_ID=294

McNulty, Sheila. (2009, June 10). Unconventional sources promise rich natural gas harvest. Retrieved February 1, 2010, from http://www.ft.com/cms/s/0/bb637bde-5556-11de-b5d4-00144feabdc0.html

Messer, Andrea, & Fong, Vicky. (2008, January 17). Unconventional natural gas reservoir could boost US supply. Retrieved February 1, 2010, from http://live.psu.edu/story/28116

Sweetnam, Glen, & US Energy Information Administration (USEIA). (2009). *International gas outlook.* Retrieved February 2, 2010, from http://csis.org/files/attachments/091028_eia_sweetnam.pdf

United States Energy Information Administration (USEIA). (1998). Natural gas 1998: Issues and trends (DOE/EIA-0560(98)). Retrieved February 16, 2010, from http://www.eia.doe.gov/oil_gas/natural_gas/analysis_publications/natural_gas_1998_issues_and_trends/it98.html

United States Energy Information Administration (USEIA). (2008a). Annual energy outlook retrospective review: Evaluations of projections in past editions (1982–2008). Retrieved March 6, 2010, from http://www.eia.doe.gov/oiaf/analysispaper/retrospective/retrospective_review.html

United States Energy Information Administration (USEIA). (2008b). Emissions of greenhouse gases in the United States 2007 (DOE/EIA-0573(2007). Retrieved February 2, 2010, from http://www.eia.doe.gov/oiaf/1605/archive/gg08rpt/index.html

United States Energy Information Administration (USEIA). (2009a). *Annual energy review 2008.* Retrieved February 2, 2010, from http://www.eia.doe.gov/aer/pdf/aer.pdf

United States Energy Information Administration (USEIA). *(2009b). Assumptions to the annual energy outlook 2009. Retrieved February 2, 2010, from* http://www.eia.doe.gov/oiaf/aeo/assumption/pdf/0554(2009).pdf

United States Energy Information Administration (USEIA). *(2009c). Table 5: US average monthly bill by sector, census division, and state 2008. Retrieved January 14, 2010, from http://www.eia.doe.gov/cneaf/electricity/esr/table5.html*

United States Energy Information Administration (USEIA). (2010a). *Electric power annual.* Retrieved February 2, 2010, from http://www.eia.doe.gov/cneaf/electricity/epa/epa.pdf

United States Energy Information Administration (USEIA). (2010b). *Electric power annual, table 2.4: Planned nameplate capacity additions from new generators, by energy source, 2008 through 2012.* Retrieved February 2, 2012, from http://www.eia.doe.gov/cneaf/electricity/epa/epat2p4.html

United States Energy Information Administration (USEIA). (2010c). *Electric power monthly, table 4.2: Receipts, average cost and quality of fossil fuels.* Retrieved February 2, 2010, from http://www.eia.doe.gov/cneaf/electricity/epm/table4_2p2.html

United States Geological Survey (USGS). (2000). *US geological survey world petroleum assessment 2000—Description and results: World assessment summaries.* Retrieved February 2, 2010, from http://pubs.usgs.gov/dds/dds-060/sum1.html#TOP

United States Geological Survey (USGS). (2003, February). *Assessment of undiscovered oil and gas resources of the Appalachian Basin Province, 2002* (USGS Fact Sheet FS-009–03). Retrieved February 1, 2010, from http://pubs.usgs.gov/fs/fs-009–03/FS-009–03-508.pdf

United States Geological Survey (USGS). (2008, October). *Assessment of gas hydrate resources on the North Slope, Alaska, 2008* (Fact Sheet 2008–3073). Retrieved February 1, 2010, from http://pubs.usgs.gov/fs/2008/3073/pdf/FS08-3073_508.pdf

United States Geological Survey. (2009). *Significant gas resource discovered in US Gulf of Mexico* [Press release]. Retrieved February 2, 2010, from http://www.usgs.gov/newsroom/article.asp?ID=2227&from=rss_home

Watson, N. J. (2009). Unconventional gas could add 60–250% to global reserves. Retrieved March 5, 2010, from http://www.petroleum-economist.com/default.asp?page=14&PubID=46&ISS=25487&SID=722880

Energy Industries—Nuclear

The nuclear power industry evolved from the same technology that developed the atomic bombs used during World War II. While some advocates champion its abundant, clean, economical, and sustainable properties, others question the safety of power plants and storage and/or disposal of spent nuclear fuel. Research and innovations in nuclear technology will continue to inspire both sides of the debate.

Particle physics was about fifty years old in January 1939 when Lise Meitner and Otto Frisch used the word *fission* to describe a new process in which an atom's nucleus splits into two parts, releasing large amounts of energy. Researchers learned that uranium isotope 235 (U-235) would produce a chain reaction and an explosion thousands of times more powerful than any chemical explosive. In February 1941, the scientist Glenn Seaborg and colleagues reported a new element—plutonium—formed when the other uranium isotope (U-238) absorbed a neutron. Plutonium fissions like uranium, but it takes less plutonium metal to make a bomb.

The United States entered World War II on 7 December 1941, and everything nuclear under development by the Manhattan District Project was classified top secret. On 6 August1945, the uranium bomb exploded over Hiroshima, and on 9 August 1945, a plutonium bomb exploded over Nagasaki, Japan. Worldwide shock and fear about "the bomb" followed this event (Rhodes 1986).

Nuclear energy was tamed when Admiral Hyman Rickover's group developed a reactor to power submarines (Bodansky 2004, 31), and the USS *Nautilus* was launched 9 November 1954. During the same time frame, industrial firms that built the US Navy reactors (including General Electric and Westinghouse) also constructed domestic electric power plants. Some of the navy's nuclear technology was declassified, and federal funds were made available to develop nuclear reactors to generate commercial utility electricity.

Nuclear Power Plants Today

Editor's note: this article predates the March 2011 tsunami that crippled Japan's Fukushima Daiichi Nuclear Power Station and left thousands of people dead and thousands more homeless. This was a watershed event for the nuclear power industry. Please see the article on "Uranium" for information on that accident.

A nuclear power plant replaces the furnace that burns coal, oil, gas, trash, or other fuels with a nuclear reactor designed to control the rate of fission to produce heat and steam that turn turbines to produce electricity. The 104 nuclear power plants in the United States, as of 2009, produce about 20 percent of the nation's electricity (WNA 2009a). The plants employ either boiling water reactors (BWRs) or pressurized water reactors (PWRs), the latter of which will be used to describe below how nuclear reactors work (Cochran and Tsoulfanidis 1990, 84–95).

Uranium from the mine is 0.71 percent U-235 and 99.29 percent U-238. The uranium metal is recovered as a "yellow cake" (U_2O_8) that is converted to uranium hexafluoride (UF_6), a solid at room temperature that melts at 64.6°C. It forms a vapor at low pressure that is passed through a series of high-speed centrifuges, which enriches the slightly lighter-weight U-235 to the PWR fuel composition of 3.75 percent U-235 and 96.25 percent U-238. This mechanical separation cannot remove all of the U-235 from mine-grade uranium. To make one tonne of enriched uranium fuel, about seven tonnes of natural uranium are required. This leaves about six tonnes of "depleted" UF_6 that contains about 0.2 percent U-235. The depleted uranium hexafluoride is not used and is stored in stainless steel cylinders (Bodansky 2004, 208).

The fuel-grade uranium hexafluoride is converted to uranium oxide powder that is pressed into pellets. The pellets are sintered (forming a solid much like fine china), and each pellet is precision ground to a diameter of 0.8 centimeters and length of 1.35 centimeters. The pellets are loaded into thin-walled alloy tubes 1.0 centimeter in diameter and 3.7 meters long, and each tube is filled with helium and sealed. Extra space in the tube allows the fuel pellets to swell because when a uranium atom fissions, it forms two fission product atoms that take a little more space. Xenon and krypton are fission product gases that collect in the fuel rod.

A 17 × 17 array of fuel rods forms a PWR fuel assembly. There are both top and bottom plates that distribute circulating water, and several spacer plates located between the end plates hold the fuel rods in a fixed position. Some tube positions hold radiation and temperature measurement probes, and there are control rods to maintain the fission rate at a constant or to shut down the reactor. A reactor core contains about 190 fuel assemblies (about 50,000 fuel rods), about 125 tonnes of the uranium oxide fuel pellets.

The core of the reactor is placed in a massive cylindrical steel pressure vessel that is about 12 meters high and 4.5 meters in diameter; the walls are 30 centimeters thick. Water remains liquid at about 15.5 MPa (2,250 psi) pressure while it is heated from about 290° to 325°C and is continuously pumped through the reactor core. (A megapascal, or MPa, is a unit of pressure measurement, as is "pounds per square inch," or psi.) The high-pressure hot water is circulated through steam generators to produce the steam that spins the turbine to produce electricity. The liquid water also serves to "slow down" the neutrons to increase the probability that fission will occur.

The nuclear reactor, water pumps, and the steam generators are located in a massive, reinforced concrete building (*reactor containment*) that is closed and sealed when the reactor is running. This isolates the reactor from the reactor operators and the public.

Each fuel assembly remains in the reactor for about three years. As the U-235 in the fuel assembly fissions, some of the U-238 captures a neutron and is converted to plutonium. The resultant plutonium is also a nuclear fuel that fissions, yielding about 40 percent of the energy produced during the three-year fuel cycle (Bodansky 2004, 212). The reactor is shut down, and new fuel assemblies replace half of the core assemblies on an eighteen-month refueling schedule.

The spent fuel elements are very radioactive and spontaneously release radiation and heat. They are stored in a pool of water containing dissolved boron, which absorbs neutrons that are released. The water collects the radioactive decay heat and serves as a shield to stop the penetrating gamma radiation (high energy X-rays). Each nuclear power plant has a storage pool designed to accumulate the spent fuel over the designed life of the power plant. Recently, the United States Nuclear Regulatory Commission (NRC) has approved applications to extend the operating license of some older nuclear plants from forty to sixty years. At these sites, "old" spent fuel is moved to dry storage containers on the power plant site.

Nuclear Power Plant Safety

Safety is a major concern in discussions about nuclear energy. Nuclear power plant safety has two components: (1) protect workers and the public from radiation and (2) minimize the risk of injury from using heavy equipment. Many occupational safety statistics have been reported over the past forty-plus years of nuclear reactor operations in the United States and the United Kingdom. The World Nuclear Association (2008) has collected data on the total number of "mine to electricity" fatalities (not including power plant construction) for the years 1970 to 1992. When each energy source produces the same amount of electricity, the number of deaths that occur for each energy source is shown in table 1.

The nuclear power industry is careful to maintain low employee exposure to radiation. In the United States, the NRC sets the limits for radiation exposure to workers. All procedures for operating the power plant reactor and handling radioactive material must comply with the NRC guidelines. Everyone (worker or visitor) must carry a radiation dosimeter while on the plant site, and the health physics staff maintains records of every individual's accumulated exposure to radiation. (*Health physics* is defined as the science of human health and radiation exposure). The NRC has authority to shut down any facility licensed to

TABLE 1. Comparison of Accident Statistics in Primary Energy Production

Energy Source	Deaths per TWy* Electricity	Who is affected?
Coal	342	Workers
Natural Gas	85	Workers and public
Hydroelectric	883	Public
Nuclear	8	Workers

*TWy is one trillion (one-million million) watts produced for a year, or 876,600 gigawatt-hours.

Source: World Nuclear Association (2008).

The number of deaths related to electricity generation in the nuclear energy industry is considerably less than the coal, natural gas, or hydroelectric industries per unit of electricity generated.

possess or handle radioactive materials that fails to observe the approved procedures.

The occupational radiation exposure to workers at nuclear installations world-wide in 2009 was about half that of exposures in 1990 (IAEA 2009, 17). Total population exposures continue to increase due to the increased use of radiation in medical procedures.

Civilian nuclear power plants have operated for fifty years—more than 12,700 cumulative reactor-years in thirty-two countries with two major accidents. Three Mile Island's reactor core meltdown near Harrisburg, Pennsylvania, in 1979 produced a financial loss, but radiation exposure was limited to the reactor containment building with essentially no health or environmental consequences (WNA 2008). In 1986 there was a steam explosion at the Chernobyl, Ukraine, reactor and a fire fueled by tonnes of graphite (pure carbon) in the reactor core, which produced widespread radiation poisoning since there was no reactor containment building. There were fifty-six deaths officially reported (although the actual number is thought to be much higher), and significant health and environmental consequences are still being monitored. These two accidents were a major blow to public confidence in nuclear power programs (Bodansky 2004, 436).

Radiation is everywhere, and at sea level exposure is about 300 millirem (mrem, an energy scale used for subatomic particles and radiation) per year (Idaho Department of Environmental Quality 2009). When proper procedures are followed, however, the radioactive burden presented to the public outside a nuclear power plant is nearly zero. The operating license requirement to protect nuclear power–plant workers and the public from radiation is a powerful incentive to maintain this low radiation exposure and should assure those who still fear nuclear power plants.

Nuclear Power Plant Economics

There are two important questions to consider when it is necessary to add any power plant to supply customer electricity: how much will the "metered" electricity cost, and how does the utility (usually an investor-owned firm) cover the power plant construction costs?

All electric power plants are expensive. It is very complicated to estimate the cost to build a nuclear power plant and the price of electricity over the forty- to sixty-year life of the plant. A panel of experts at Massachusetts Institute of Technology (2009) estimated the costs to build fossil fuel–fired and nuclear power plants, shown in table 2.

The "overnight cost" is the capital investment in dollars per kilowatt installed capacity ($/kW) that it will cost to build the power plant, assuming all funds are spent in one day (no site development, construction delay, or construction loan costs are included). The "base case" (engineering estimates based on today's "best available" power plant technology for each energy source) assumes nuclear power plant investments are risky, requiring 10 percent interest charge; interest for coal and natural gas are set at 7.8 percent. When the nuclear plant interest is dropped to an equivalent 7.8 percent, the estimate shows nuclear power is cost competitive with coal and natural gas "at the meter." The only real test of these estimates is to build the power plants and sell power over the years the plant operates. Most states within the United States have a public service commission that approves any plan to build a power plant, and they review (usually annually or when the utility requests a rate increase) the retail price of the power, a control on "free market" excesses.

Available Uranium for Nuclear Fuel

All new nuclear power plants built in the next thirty years will be water-cooled reactors based on the best available technology; any new reactor technology will not be ready to deploy for about thirty years. Reactors are designed to use a "once through" uranium fuel cycle that extracts only 3–4 percent of the energy available in the fuel. Spent fuel is removed, stored, and replaced by new fuel.

It takes about seven tonnes of natural uranium (0.71 percent U-235) to produce one tonne of pressurized water

Table 2. **Cost of Electric Generation Alternatives**

Plant type	Overnight Cost ($/kW)	Fuel Cost ($/million Btu)	Base Case Cost (¢/kWh)	w/carbon charge $25/tonne CO_2 (¢/kWh)	w/same interest ¢/kWh
Nuclear	4,000	0.67	8.4		6.6
Coal	2,300	2.60	6.2	8.3	—
Natural Gas	850	7.00	6.5	7.4	—

Source: Massachusetts Institute of Technology (2009, 6).

Although electricity generated by coal and natural gas is less costly when the perceived higher "risk" of nuclear energy is considered (base case), nuclear power is actually competitive with other energy sources when the assumed risks are equivalent.

reactor (PWR) fuel (3.75 percent U-235). This will leave about six tonnes of depleted uranium (0.2 percent U-235). Fuel enrichment in the United States has produced about 480,000 tonnes of heavy metal (MTHM) of depleted uranium; the figure worldwide is about 1,189,000 MTHM (Depleted Uranium 2008). Each year the power plants add about 12,600 MTHM to the US inventory.

A report prepared for Congress stated that the US inventory of spent nuclear fuel in 2002 was approximately 47,000 MTHM, and every year the 104 operating commercial reactors contribute an additional 2,150 MTHM of spent fuel (Andrews 2004, 3–5). Based on these figures, the current US inventory of spent nuclear fuel is estimated at approximately 64,200 MTHM.

Current spent fuel inventories present a very long-term disposal problem and a serious objection by those opposed to nuclear power. The spent fuel remains radioactive for tens of thousands of years because it contains the heavy metals neptunium, americium, and curium in addition to uranium and plutonium. It is possible to chemically separate all five elements, leaving the fission products for disposal (Bodansky 2004, 213–222). All of the elements beyond uranium in the periodic table (including all of the heavy metals listed above) become fuel in next-generation fast neutron reactors that are being developed (Bodansky 2004, 186–190).

Nuclear Reactors Worldwide

As of 2009 thirty-one countries with 436 nuclear reactors operate with an installed capacity of about 370 gigawatts (GW). One gigawatt is equal to one million watts and is enough to power 780,000 average US homes in a given instant (calculation based on data from USEIA 2009). (Please refer to the sidebar "What's the difference between a gigawatt and a gigawatt-hour?" on pages 4–5 for more on this topic.) There are fifteen countries that have fifty-three plants under construction with nameplate capacity (maximum rated output) of 47 GW. Table 3 on page 74 lists the countries with substantial nuclear power in operation and with new reactors under construction (European Nuclear Society 2010; WNA 2010). The new plants should become operational during the period 2015–2017. The new reactor construction initiative is clearly outside the United States.

The Future

All future energy options for the United States require a long-term (thirty to fifty years) politically bipartisan energy policy that provides funds for research and allows operating-plant scale comparisons of all energy options to produce sustainable commercial electricity. The safety and competitive cost of nuclear power have been demonstrated with the global fleet of reactors operating today. Since the 1980s when the United States stopped developing nuclear energy systems, growth has moved global. (See table 3.) Nuclear energy development into the next century must adhere to the following strategic components (Lister and Rosner 2009):

- The energy produced must be "cost effective."
- Safety must be a primary design objective.
- Design must minimize risk of nuclear theft and terrorism (weapons proliferation).
- Size must be appropriate to match the national distribution grid.
- New systems development should be "evolutionary" rather than "radical."

The BWR and PWR nuclear reactor technology "fine tuning" during the past twenty years has been done outside the United States. These are the reactors that will be added to the worldwide nuclear fleet in the next thirty years.

Nuclear energy becomes sustainable in the long-term when the "whole tonne" of uranium is used as fuel. Opponents of nuclear energy contend that spent nuclear fuel inventories are especially risky, thus work on the chemistry to separate the heavy metal fuel values from the fission products should continue. This recycled fuel will be a mixture of fissionable metals that cannot be used as a nuclear weapon. The Experimental Breeder Reactor II (operated from 1965 to 1995) is an example of a next-generation reactor that can use reprocessed fuel. In addition, all the U-238 in depleted uranium can be used to "close the nuclear fuel cycle" (Stacey 2007, 244). This strategy produces roughly sixty times as much energy per tonne of natural uranium as the "once through" PWR (MacKay 2009, 162).

Early in the development of domestic nuclear energy, there was concern about a shortage of uranium. Thorium was considered to be a good nuclear fuel because there is over three times as much thorium as uranium in the Earth's crust (Lide 2005, 14–17). Natural thorium is pure Th-232, so isotope enrichment that is required for PWR uranium fuel is unnecessary. When Th-232 is placed in a reactor and a thorium atom nucleus accepts a neutron, in a short time it becomes U-233 that will fission just like U-235; a thorium fuel cycle has been described (Benedict, Pigford, and Levi 1981, chapter 6). The thorium fuel option supplementing uranium

TABLE 3. **Nuclear Reactors Worldwide**

Country	Reactors In Operation	Current Output (GW)	Reactors Under Construction	Additional Output (GW)
China	11	8.4	16	15.2
France	59	63.2	1	1.6
Germany	17	20.5	0	0
India	17	20.5	6	2.9
Japan	53	46	2	2.2
Russian Federation	31	21.7	9	6.9
South Korea	20	17.6	6	6.5
Ukraine	15	13.1	2	1.9
United Kingdom	19	10.1	0	0
United States	104	100.7	1	1.1
All Others	90	48.3	10	8.8
Total	436	370.2	53	47.2

Source: European Nuclear Society (2010).

Construction of new nuclear power facilities for 2010 and beyond is concentrated in countries outside the United States, currently the world leader in nuclear power generation output; China, currently the world's largest consumer of coal, has the most nuclear power plants planned.

fuel can supply the growing world energy demand for thousands of years (MacKay 2009, 166).

Sustainable energy options into the future will include solar, wind, biofuels, and nuclear. Today nuclear power plants provide 20 percent of the electricity in the United States and about 16 percent globally using incremental improvements on 1960s technology. Sustainability depends on closing the nuclear fuel cycle. The chemistry and physics required to develop "new reactor technology" and use these fuels has been identified. But science and technology is fourth on the list: there must be the national political will, funding must be provided, and public opinion must be supportive before new nuclear technology will be demonstrated.

The energy infrastructure is huge, and additions to meet demands will be expensive, so any changes will be incremental. This provides time to analyze all of the energy options, choose among sustainable energy paths, conduct the science-based research, and demonstrate commercial-scale advanced technologies (Peters 2009). Nuclear energy is about fifty years old, and the developments described here could take another one hundred years. That is enough time.

Truman STORVICK
University of Missouri, Emeritus

See also in the *Berkshire Encyclopedia of Sustainability* Consumer Behavior; Energy Efficiency; Energy Industries—Overview of Renewables; Mining; True Cost Economics

FURTHER READING

Andrews, Anthony. (2004, December 21). Spent nuclear fuel storage locations and inventory (CRS Report to Congress). Retrieved October 30, 2009, from http://ncseonline.org/NLE/CRSreports/04Dec/RS22001.pdf

Benedict, Manson; Pigford, Thomas H., & Levi, Hans Wolfgang. (1981). *Nuclear chemical engineering*, (2nd ed.) New York: McGraw-Hill.

Bodansky, David. (2004). *Nuclear energy: Principles, practice, and prospects* (2nd ed.). New York: Springer-Verlag.

Cochran, Robert G., & Tsoulfanidis, Nicholas. (1990). *The nuclear fuel cycle: Analysis and management (2nd ed.).* LaGrange Park, IL: American Nuclear Society.

Depleted Uranium Inventories (2008, April 21). Retrieved October 30, 2009, from www.wise-uranium.org/eddat.html

European Nuclear Society. (2010). Nuclear power plants world-wide. Retrieved October 14, 2009, from http://www.euronuclear.org/info/encyclopedia/n/nuclear-power-plant-world-wide.htm

Idaho Department of Environmental Quality. (2009). INL oversight program: Guide to radiation doses and limits. Retrieved October 15, 2009, from http://www.deq.state.id.us/inl_oversight/radiation/radiation_guide.cfm

International Atomic Energy Agency (IAEA). (2009). *Nuclear safety review for the year 2008*. Retrieved November 5, 2009, from http://www.iaea.org/About/Policy/GC/GC53/GC53InfDocuments/English/gc53inf-2_en.pdf

Lide, David R. (2005). *CRC Handbook of Chemistry and Physics* (86th ed.). Oxford, UK: Taylor and Francis.

Lister, Richard K., & Rosner, Robert. (2009). The growth of nuclear power: Drivers & constraints. *Daedalus, 138*(4), 19–30.

MacKay, David J. C. (2009). *Sustainable energy: Without the hot air.* Cambridge, UK: UIT Cambridge.

Meitner, Lise, & Frisch, Otto R. (1939). Disintegration of uranium by neutrons: A new type of nuclear reaction. *Nature, 143*, 239–240.

Massachusetts Institute of Technology. (2009). *Update of the MIT 2003 future of nuclear power study: An interdisciplinary MIT study.* Retrieved October 15, 2009, from http://web.mit.edu/nuclearpower/pdf/nuclearpower-update2009.pdf

Peters, Mark T. (2009, June 17). Testimony to US House Committee on Science and Technology on advanced nuclear fuel cycle research and development. Retrieved December 2, 2009 from http://www.anl.gov/Media_Center/News/2009/testimony090617.pdf

Rhodes, Richard. (1986). *The making of the atomic bomb.* New York: Simon & Schuster.

Seaborg, Glenn T. (1972). *Nuclear milestones: A collection of speeches.* San Francisco: W. H. Freeman.

Stacey, Weston M. (2007). *Nuclear reactor physics* (2nd ed.). Weinheim, Germany: Wiley-VCH GmbH.

United States Energy Information Administration (USEIA). (2009). Table 5: US average monthly bill by sector, census division, and state 2007. Retrieved January 14, 2010, from http://www.eia.doe.gov/cneaf/electricity/esr/table5.html

World Nuclear Association (WNA). (2008) Safety of nuclear power reactors. Retrieved February 10, 2010, from http://www.world-nuclear.org/info/inf06.html

World Nuclear Association (WNA). (2009). Nuclear power in the USA. Retrieved January 15, 2010, from http://www.world-nuclear.org/info/inf41.html

World Nuclear Association (WNA). (2010). Plans for new reactors worldwide. Retrieved January 15, 2010, from http://www.world-nuclear.org/info/inf17.htm

Energy Industries—Solar

With continuing advances in technology, solar power has the potential to become an important source of renewable energy. The vast majority currently comes from solar thermal systems that use the sun's energy to heat water either for direct use or to generate electricity. Less solar power results from photovoltaic technology, which produces electricity from specific light frequencies of the sun.

In a given hour, the Earth receives more energy from the sun than humanity uses in a year. Only a fraction of this energy is usable due to limits in available surface area and technology inefficiency, yet solar power has the potential to meet much of humanity's energy needs. At the beginning of the twenty-first century, solar power is used in handheld devices, toys, hiking and camping equipment, street and landscape lighting, utility-scale solar power plants, and much more. As solar technology and manufacturing continue to advance, solar power has the capability to become an increasingly integral part of society.

Solar power comes in a variety of forms, and solar technologies can be categorized and compared in several ways, including solar thermal versus photovoltaic and concentrated solar versus flat panel solar.

Solar Thermal Technologies

Solar thermal systems use sunlight for heat, which can be used to provide warm water, to produce electricity, or for other applications. According to the policy network REN21 (2009, 9), approximately 145 gigawatts (GW) of solar thermal capacity have been installed globally, which represents more than 90 percent of total installed solar capacity (see table 1 on page 77). (A watt is a measure of power, the *rate* at which energy is converted to electricity. One gigawatt is equal to one billion watts and is enough to power 780,000 average US homes in a given instant [calculation based on data from USEIA 2009b]. For more on this topic please refer to the sidebar "What's the difference between a gigawatt and a gigawatt-hour?" on pages 4–5.)

A common solar-thermal configuration consists of enclosed tubes filled with a heat transfer fluid (HTF). Sunlight strikes the tubes and warms the HTF, which in turn heats a water supply. Solar hot-water systems can function in diffuse sunlight and below the freezing point of water. They reduce the need for heating water by other means, and these systems constitute more than 145 GW of installed capacity, providing hot water for tens of millions of people (REN21 2009, 9).

Other solar thermal systems use heat to produce electricity and are often referred to as concentrated solar power (CSP). In some CSP systems, reflective surfaces are arranged to concentrate sunlight, which warms an HTF. In a trough system, a parabolic trough of reflective material focuses sunlight on a coaxial tube of oil. In a tower system, reflective surfaces are arranged over a large area to focus sunlight onto a tower, perhaps filled with molten salt. In both trough and tower systems, the HTF carries heat to a power generator, such as a steam turbine. Concentrating thermal systems can be used as independent sources of electricity, or they can be used to augment heating at coal or natural gas plants to reduce fossil fuel use.

There are numerous advantages for CSP. Because heat dissipates slowly from an HTF reservoir, system operators can capture heat throughout the day and then use the captured heat to produce electricity based on consumer demand. These systems require relatively inexpensive materials, which reduce costs, and technology for thermal power

TABLE 1. **Key Statistics**

Technology	Installed Capacity (December 2008)	Capacity CAGR* (2002–2008)	Market Potential, All Solar Technologies (% of Global Energy Use)
Thermal	146 GW		
Water Heating	145 GW	17%	Estimates vary widely. UN: 30% by 2040 USEIA**: 2% by 2030 ExxonMobil: <1% by 2030
Concentrated	0.5 GW		
Photovoltaic	13 GW		
Flat-Panel	13 GW	55%	
Concentrated	< 0.1 GW		

* CAGR = Compound Annual Growth Rate

** USEIA = United States Energy Information Administration

Sources: Installed capacity figures from REN21 (2009, 23); CAGR figures from REN21 (2009, 9) and REN21 (2005, 9); market potential figures from Resch and Kaye (2007, 63), USEIA (2009a, 109), ExxonMobil (2008, 38).

Most installed solar capacity (almost 90 percent) uses solar thermal (heating) systems, although photovoltaic technologies saw a greater percentage of growth from 2002 to 2008.

generation is well understood. Concentrating thermal systems currently account for less than 1 GW of installed global capacity, although the technology's advantages make it a promising future source of energy.

Solar thermal systems are not without disadvantages, however. In order to operate efficiently, each reflective surface must accurately follow the path of the sun. Solar thermal systems perform best in continuous, direct sunlight, which makes CSP impractical for many parts of the world. Solar thermal systems require water for cooling and for cleaning the reflective surfaces. Existing CSP plants are located in low-latitude deserts to utilize the abundant sunshine, but water is a limited resource in the desert.

Not all solar thermal systems use an HTF. Parabolic dishes can be used to concentrate sunlight as a heat source for a Stirling engine, a closed-system engine in which a working gas (typically air, helium, or hydrogen) is alternately heated (expanded) and cooled (contracted) to push and pull the piston and generate electricity. A Stirling system benefits from components that are very cost-effective, and it does not require an expensive HTF. Because there is no special fluid or separate steam turbine, a Stirling system may involve simpler engineering than some CSP technologies. Finally a Stirling system may be more scalable than other solar thermal systems because a field of Stirling dishes can be expanded simply by adding more dishes.

Solar thermal power can also be used in what is called a solar convection tower, or solar updraft tower. In such a system, a large area is covered by a greenhouse with a tall chimney in the middle. Sunlight warms the air in the greenhouse, and the warmed air is drawn to the middle of the structure in order to rise through the chimney. One or more turbines are located at the base of the chimney or in the chimney itself, and as the air rushes by them, the turbines produce electricity.

Solar updraft towers use inexpensive and plentiful materials and enable dual-use of land: the greenhouse retains moisture in the air and soil, which can enable agriculture in the structure. Providing both electricity and food could make a solar updraft tower appealing in an arid climate. A utility-scale facility would need to cover hundreds of hectares, however, and would need a chimney hundreds of meters tall.

Photovoltaic Technologies

Photovoltaic (PV) materials produce direct current (DC) electricity when exposed to light of a specific frequency. When a photon strikes a PV material, the photon can excite an electron and liberate it from its atom. In a PV cell, one side of the semiconductor is n-type (with an abundance of free electrons) and the other is p-type (with an abundance of available holes for electrons). Creating n-type and p-type semiconductors can be achieved by adding impurities such as phosphorous or boron. When n-type and p-type materials are brought together, an electric field forms. The electric field provides resistance for electrons moving from the n-type to the p-type, and it encourages electrons to move from the p-type to the n-type. The result is that incoming sunlight produces a steady build-up of positive charge on one side of the material and a negative charge on the other. Proper wiring creates a circuit with an electric current.

Photovoltaic materials are often categorized as either crystalline silicon (c-Si) or thin film. Crystalline silicon includes both monocrystalline and polycrystalline technology. Thin film includes a variety of materials, including amorphous silicon (a-Si), cadmium telluride (CdTe), and copper indium gallium selenide (CIGS). The PV market has averaged an annual growth rate of approximately 55 percent from 2004 to 2008, with both c-Si and thin film segments growing rapidly (REN21 2009, 9; REN21 2005, 9).

Crystalline silicon's high efficiency makes it an appealing technology where area is at a premium, such as on satellites and rooftops. Thin film PV material and manufacturing costs may be substantially cheaper than those for c-Si, although thin film PV is less efficient at converting sunlight into electricity. As a result, thin film is most common in large ground-mounted arrays, such as utility-scale power plants.

In the second half of the twentieth century, silicon was widely used in electronics because of its semiconductor properties. Much of the silicon used early in the PV industry was waste material from the semiconductor industry, which reduced the cost of the primary PV material. Refining and manufacturing processes remain costly, and silicon accounts for as much as half of the total cost of a c-Si PV system.

To produce monocrystalline silicon, raw silicon is purified using chemical processes and then melted. The molten silicon is allowed to cool slowly while in contact with a single crystal, called a "seed." The result is an ingot of very pure silicon, which is then sliced into thin wafers no more than a few millimeters thick. These wafers are treated to add the desired amount of specific impurities to produce n-type and p-type silicon.

An alternate method of producing silicon cells is using a process referred to as "ribbon growth." In this process, molten silicon forms a sheet as it is passed between two seed crystals. Ribbon growth dramatically reduces waste that results from slicing ingots into wafers, although wafers from ribbon growth are generally lower quality than those from ingots.

Producing polycrystalline silicon often involves casting molten silicon into a mold where the silicon cools to form an ingot. Polycrystalline wafers are less costly to produce than monocrystalline wafers because the process is simpler and can make use of lower quality feedstock. The trade-off is that polycrystalline wafers are less efficient than monocrystalline.

Completed wafers make solar cells, which are grouped to form a module (sometimes called a "panel") that has a protective coating and conductive strips to carry electrical current away from each cell. Modules are arranged to make a string, and strings are combined to form a solar array, which are often seen on a rooftop.

Thin film does not require wafers and cells; instead, it uses homogenous layers of PV material deposited on a substrate, or backing. There are multiple options for the deposition method, the PV material, and the substrate, which creates a great deal of diversity within the thin film market. Of particular interest for thin film is the ability to employ a flexible backing material, such as plastic or thin metal. The resulting PV material could be unrolled quickly onto a surface and could be shaped to fit curved surfaces.

Amorphous silicon was the first thin film technology used commercially and remains a commonly used material. Another thin film PV material is cadmium telluride (CdTe), which has become increasingly common in the twenty-first century. Compared to c-Si, CdTe has gained ground because it is cheaper to manufacture; from 2006 to 2008 in particular, crystalline supply issues caused c-Si prices to be both high and volatile. More recently, copper indium gallium selenide (CIGS) technology has emerged as another alternative, and experiments indicate that CIGS may be able to attain higher efficiencies than CdTe. (Efficiency is a measure of how much electricity is produced by a given amount of sunlight.) But CdTe has benefited from being first to market: CIGS and CdTe use similar manufacturing methods, and so both profit from low marginal costs, but the first thin film company to establish high-volume, high-quality product used CdTe.

Sunlight contains a broad electromagnetic spectrum, but a given material is only photovoltaic for a limited band of the spectrum. One way to improve solar PV efficiency is by using multijunction cells, which are essentially layers of PV materials. Each material layer captures photons of a different frequency.

Producing multijunction cells is a complicated and costly process, which makes these cells substantially more expensive than other PV materials. One important application for multijunction cells is in concentrating PV (CPV) systems. CPV systems use lenses or reflective surfaces to focus sunlight on a small area of high-efficiency PV material. CPV systems require far less PV material than do other PV systems. This reduction in material reduces system cost and can make an expensive, high-efficiency PV material worthwhile. But CPV systems are ill suited for many regions because concentrating technology requires continuous, direct sunlight. CPV systems require precise tracking mechanisms to follow the sun's daily and seasonal movements. In addition, extreme temperatures lower the PV efficiency and risk damaging the PV cells, so particular care must be taken to cool the PV material.

Photovoltaic technologies offer ample efficiency and cost combinations to meet the needs of a given application. For single-junction monocrystalline cells, laboratory experiments have achieved efficiencies of approximately 25 percent; production-grade cells can operate at efficiencies as high as 20 percent; lower quality c-Si cells perform below 15 percent efficiency. Laboratory experiments with thin films have achieved efficiencies near 20 percent. The highest-efficiency thin films commercially available operate above 10 percent, and less costly thin film is available with efficiency below 8 percent. Multijunction cells have surpassed 40 percent efficiency in laboratory experiments, and production-line cells are available with efficiencies above 35 percent.

An advantage of PV materials, particularly thin films, is that they function in both direct and diffuse sunlight, and new materials may utilize infrared or other frequencies. A PV system can be tailored to specific power needs and expanded as needs change. Constructing a large PV system can take months, while other technologies can require years to develop a comparable size system. A key disadvantage of PV materials is the constant need for sunlight, which makes changing cloud cover an issue and nighttime operation impractical. Another drawback for PV systems is cost. Improvements in materials and manufacturing processes have resulted in substantial cost reductions, but PV systems remain a costly source of electricity.

The Future of Solar Power

Concentrated solar technology continues to advance, which has led to increased investment in CSP projects: CSP accounts for less than 5 percent of global solar power capacity, but CSP accounts for more than a quarter of planned solar power projects. With further improvements in technology, CSP could provide an on-demand source of electricity, and CSP may become a viable source of electricity for ever-broader regions. Continued cost reductions could make CSP increasingly cost-competitive with other forms of utility-scale electricity. Thermal storage systems may allow CSP to provide power from sunset one day until sunrise the next.

Solar PV technology is often divided into three "generations," and by most accounts, the industry is completing its transition from the first to the second generation. The first generation involved the most basic materials and technology, such as selenium and silicon, and solar cells were very costly. The c-Si cells in use today are still considered first generation, although efficiencies and manufacturing processes have improved substantially over the last fifty years.

Second generation technology is marked by improvements in materials, efficiencies, and costs. Multijunction cells and thin film solar both represent advances into the second generation of solar PV with improved cost per energy output.

Third generation solar cells will be characterized by further advances in materials and manufacturing and may approach properties of photosynthetic plants. Graphene (a single sheet of graphite only one atom thick) or quantum dot (another very thin semiconductor) technologies may serve as solar semiconductors, or semiconductors may be eliminated by advances with polymers (synthetics) or organics. In a world of third-generation solar technology, PV devices may be as common as cloth or paint are today, and any man-made surface could produce electricity.

An alternate approach would use orbiting satellites as large solar arrays, transmitting energy to the surface of the Earth using low-intensity microwave radiation. The most significant advantage for this technology would be the ability to produce electricity at all times of day in all weather conditions. Solar power transmitted from satellites may seem like science fiction, but serious research is underway. In early 2009, a California utility company announced an agreement to purchase electricity from just such a space-based solar array beginning in 2016 (Riddell 2009). Later in 2009, a consortium of Japanese manufacturing companies announced plans to develop the technology for large-scale application (Sato and Okada 2009).

Innovation and technological advances will continue to play an important role in the development of solar power. New materials, improved manufacturing processes, and creative ways of using solar power will all be critical. Solar power has the potential to become an integral component of energy production for human civilization.

Michael Dale HARKNESS
Cornell University

See also in the *Berkshire Encyclopedia of Sustainability* Cap-and-Trade Legislation; Development, Sustainable; Energy Efficiency; Energy Industries—Overview of Renewables; Facilities Management; Investment, CleanTech

Further Reading

American Solar Energy Society. (2009). Retrieved September 2, 2009, from http://www.ases.org/

Erwing, Rex A. (2006). *Power with nature: Alternative energy solutions for homeowners* (2nd ed.). Masonville, CO: PixyJack Press.

ExxonMobil. (2008). The outlook for energy: A view to 2030. Retrieved November 18, 2009, from http://www.exxonmobil.com/corporate/files/news_pub_2008_energyoutlook.pdf

International Energy Agency. (2009). Renewable Energy. Retrieved September 2, 2009, from http://iea.org/Textbase/subjectqueries/keyresult.asp?KEYWORD_ID=4116

International Solar Energy Society. (2009). Retrieved September 2, 2009, from http://www.ises.org/

Marshall, J. M., & Dimova-Malinovska, D. (Eds.). (2002). *Photovoltaic and photoactive materials: Properties, technology, and applications*. Boston: Kluwer Academic.

Martin, Christopher L., & Goswami, D. Yogi. (2005). *Solar energy pocket reference*. London: Earthscan.

Martinot, Eric; REN21; & Worldwatch Institute. (2007). Renewables 2007: Global status report. Retrieved September 2, 2009, from http://www.worldwatch.org/files/pdf/renewables2007.pdf

Masters, Gilbert M. (2004). *Renewable and efficient electric power systems*. Hoboken, NJ: John Wiley & Sons.

National Renewable Energy Laboratory. (2001, March). Concentrating solar power: Energy from mirrors. Retrieved September 2, 2009, from http://www.nrel.gov/docs/fy01osti/28751.pdf

National Renewable Energy Laboratory. (2009). Solar research. Retrieved October 27, 2009, from http://www.nrel.gov/solar/

Navigant Consulting. (2009). *Solar outlook* [Newsletter]. Retrieved September 2, 2009, from http://www.navigantconsulting.com/industries/energy/renewable_energy/solar_outlook_newsletter/

Patel, Mukund R. (2006). *Wind and solar power systems: Design, analysis, and operation* (2nd ed.). Boca Raton, FL: Taylor & Francis

REN21. (2005). *Renewables 2005: Global status report*. Retrieved November 18, 2009, from http://www.ren21.net/pdf/RE2005_Global_Status_Report.pdf

REN21. (2009). *Renewables global status report: 2009 update*. Retrieved November 18, 2009, from http://www.ren21.net/pdf/RE_GSR_2009_Update.pdf

Renewable Energy World. (2009). Solar energy. Retrieved September 2, 2009, from http://www.renewableenergyworld.com/rea/tech/solarenergy

Resch, Rhone, & Kaye, Noah. (2007, February). The promise of solar energy: A low-carbon energy strategy for the 21st century. Retrieved November 18, 2009, from http://www.un.org/wcm/content/site/chronicle/cache/bypass/lang/en/home/archive/issues2007/pid/4837?ctnscroll_articleContainerList=1_0&ctnlistpagination_articleContainerList=true

Riddell, Lindsay. (2009, April 13). PG & E to source solar from outer space. Retrieved November 18, 2009, from http://sanfrancisco.bizjournals.com/sanfrancisco/stories/2009/04/13/daily10.html

Sato, Shigeru, & Okada, Yuji. (2009, September 1). Mitsubishi, IHI to join $21 Bln space solar project (update1). Retrieved November 18, 2009, from http://www.bloomberg.com/apps/news?pid=20601101&sid=aJ529lsdk9HI

Scientific American. (2009). Alternative energy technology. Retrieved September 2, 2009, from http://www.scientificamerican.com/topic.cfm?id=alternative-energy-technology

Solar Electric Power Association. (2009). Retrieved September 2, 2009, from http://www.solarelectricpower.org/

Solar Energy Industries Association. (2009). Retrieved September 2, 2009, from http://www.seia.org/

Solar Industry. (2009). Retrieved September 2, 2009, from http://www.solarindustrymag.com/

Solar Today. (2009). Retrieved September 2, 2009, from http://www.solartoday.org/

US Department of Energy / Energy Efficiency & Renewable Energy. (2009). Solar energy technologies program. Retrieved September 2, 2009, from http://www1.eere.energy.gov/solar/

US Department of Energy / Energy Information Administration. (2009). Renewable & alternative fuels. Retrieved September 2, 2009, from http://www.eia.doe.gov/fuelrenewable.html

US Energy Information Administration (USEIA). (2009a). Appendix A. In Annual energy outlook 2009 (p. 109). Retrieved November 18, 2009, from http://www.eia.doe.gov/oiaf/aeo/pdf/appa.pdf

United States Energy Information Administration (USEIA). (2009b). Table 5: US average monthly bill by sector, census division, and state 2007. Retrieved January 14, 2010, from http://www.eia.doe.gov/cneaf/electricity/esr/table5.html

Willett, Edward. (2005). *The basics of quantum physics: Understanding the photoelectric effect and line spectra*. New York: Rosen Publishing Group.

Energy Industries—Wave and Tidal

In the mid-twentieth century researchers began examining how to develop cost-effective methods for generating energy from tides and waves. Marine energy comes from the water's motion, which is caused by the wind and by gravitational interactions between the water, Earth, its moon, and the sun. Whether the various technologies, methods, and devices can efficiently generate energy from the water remains unproven.

It is obvious that there is great power in the oceans and the seas of the world. What is not so clear is whether such energy can be harnessed in a cost-effective manner. The prospect of capturing energy from the motion of the waves and the changes in the tides has received considerable interest. It is difficult to quantify exactly what contribution wave and tidal energy could make to sustainable energy generation. One current tidal energy project generated 1 gigawatt (GW) within the first five years of introducing a commercial-scale device. (A watt is a measure of power, the *rate* at which energy is converted to electricity; one gigawatt is equal to one billion watts and is enough to power 780,000 average US homes in a given instant [calculation based on data from USEIA 2009]. For more on this topic please refer to the sidebar "What's the difference between a gigawatt and a gigawatt-hour?" on pages 4–5.) Locations such as the Korean peninsula, New Zealand, the United Kingdom, and the northeastern coast of the United States all offer considerable potential. Likewise, wave energy can be generated on any coast that faces a prevailing wind and has a large expanse of ocean. Estimates for the world's total coastal wave power potential indicate that it could be as high as 1 terawatt, or 1,000 GW (Falnes 2007), which is similar to that of all existing power stations.

Tidal energy has been used for centuries to grind flour. In the mid-twentieth century, researchers conducted studies on the use of tidal barrages to generate electricity, and a number of schemes were implemented worldwide, notably the 240 megawatt (MW) barrage at La Rance in northern Brittany, France. (Tidal barrages span estuaries and tidal channels; their power plants generate electricity as the tide flows in and out.) Since 2000, research has also been done in the use of tidal currents to generate electricity. Significant wave energy extraction research was initiated following the oil supply scares of the early 1970s, although as yet there are no significant schemes generating commercial electricity.

Sources of Marine Energy

Two distinct mechanisms generate the motion of the oceans and offer the prospect of renewable sources of environmental energy. The first is the movement of the wind. This arises from the solar radiation, internal heat, and the rotation of the Earth. As the resultant wind blows over the surface of the sea, a small fraction of its energy is transferred to the water. The longer the distance that the wind blows over the sea, the greater the energy transferred. The waves generated over longer distances will also have larger amplitudes (or maximum heights of the wave crests) with a wider spectrum of wavelengths (distance between wave crests). These waves travel long distances and are reflected, radiated, and refracted by the barrier of land and the changes in sea depth. As a result, any given location in the wave environment, or sea state, will contain a spectrum of differing wavelengths. The larger the wave amplitude of any given wave, the greater its energy content.

The challenge of wave energy extraction is to "tune" the wave energy device to best capture that energy. In general the best locations for wave energy are near land where the ocean remains deep and the prevailing wind can transfer energy for thousands of kilometers. Good examples are the western coasts of Spain and Portugal and the Outer Hebrides off the northwest coast of Scotland.

Although tidal energy is also a form of wave motion, the origin of the energy is in the complex gravitational interaction between the Earth, moon, sun, and the water in the oceans. The water is effectively "pulled" by these gravitational forces, transferring energy into the sloshing of the oceans. The regularity of this second mechanism for generating motion is a key benefit for energy production. At any given location, the tidal height (or the corresponding current) is made up of the twice-daily maximum (or high) tide. Tidal height is moderated by the fortnightly variations that give rise to the so-called spring tides (which have the greatest range between high- and low-tide levels) and the neap tides (which have the lowest range between the tide levels). These can be predicted with great accuracy and they ensure a known amount of this energy supply.

Tidal energy is at a low density in the deep ocean, and thus a suitable combination of coastline or islands and a variation of sea depth are required to create a large tidal range and/or strong currents. As the time of high tide varies along a given coastline, this lag can be used to further smooth the generation of electricity if a series of sites are connected to a regional electrical grid. The difference in energy available between spring and neap tides, however, can typically result in an eightfold reduction in power.

Wave Energy Devices

The successful generation of viable amounts of electrical power has proved elusive, despite forty years of evolving technology for capturing wave energy. The devices range from concepts such as the Scottish professor Stephen Salter's "nodding duck" to the snakelike Pelamis wave energy converter and the even-more-radical deforming Anaconda rubber tube. Nevertheless there are two basic concepts for wave energy devices:

1. Waves break over an enclosed floating or fixed barrier, thereby generating a difference in water height. This difference in water height, or head, can then be used to drive a low head turbine connected to an electrical generator.
2. An oscillatory device, comparable to the weight at the end of a pendulum, responds to the periodic movement of the waves, and its motion is used to pump hydraulic fluid or to directly move a linear induction generator. An example is the Pelamis device, which consists of a series of interconnected cylinders aligned with the wave direction. The flexing of the joints between cylinders is used to pump hydraulic fluid to drive a turbine.

Many related variants on these systems exist, including a cliff-mounted chamber that uses the motion of the sea's surface to force air continually in and out of a suitable turbine. Operators of such systems have had very little experience under real conditions and at production-scale generation. As a result, the suitability of these systems for cost-effective power generation is difficult to assess.

Tidal Energy Systems

Two physical principles exist in harnessing the energy of waves: capturing the tidal range to generate a suitable water head to drive a turbine-generator set, or using the tidal current itself to directly drive a turbine.

The first principle is used in tidal barrage schemes. Usually these schemes have a barrier built across an estuary. It can generate power on the incoming, the outgoing, or both cycles of the tide. The combination of sluices and turbines limits the flow rate and thereby ensures a sufficient head of water to drive the turbine. These schemes require a large investment in the civil works needed to build the barrier and will alter the marine environment on both sides of the barrage and for a long distance upriver. But there are advantages: a large amount of power is generated at predictable times, and, if the turbines can be used as pumps, the scheme can be used for energy storage to regularize the power from more variable sources of renewable energy, such as wind.

The use of the tidal currents, or other ocean circulation currents, is relatively new. Device development has been rapid, with the demonstration systems installed from 2003 onward. The largest single device generated 1.2 MW in winter 2008 in Strangford Lough, Northern Ireland. The systems are based around wind turbine concepts that are revised for use in the sea. The successful devices have blades that rotate around a horizontal axis that is aligned with the main tidal direction. Ideally the turbines need to be able to generate power for both directions of tidal flow by either employing a yaw (side to side) mechanism or bidirectional blades. Variations use an accelerating duct to capture more flow, but the disadvantages are that the extra energy generated needs to overcome the increased structural costs and that the design of the duct must not choke the flow through the turbine. When many such systems are located as an array or fence, a significant change in the local flow will determine the maximum energy extraction.

Future Challenges

Wave energy and tidal energy systems face many barriers before it can be determined whether such schemes can extract energy at a large enough scale to make a major contribution to a more sustainable energy future. The main barrier is an economic one, although, by necessity, the installation and operation of such schemes will affect their local environment. A considerable challenge is assessing whether this impact is significant when set against the benefits of using environmental energy.

Tidal barrage schemes require large investment with a payback period running into many decades; they also cause a significant change in the estuarine environment. They are relatively low risk as they use tried and trusted technology.

Wave energy is still an unproven technology. Designing devices that can survive the harshest storms and yet still be cost-effective is only one difficulty to surmount.

Tidal current systems benefit from being submerged away from the worst wave loadings, but that comes at the cost of difficult access for installation and maintenance. It is likely that these systems will improve progressively to become a viable alternative source of renewable energy. Unfortunately this commercialization phase is the most expensive part of the development cycle and requires the accumulation of many years of operational experience before cost-effective and reliable systems can be installed.

Stephen TURNOCK
University of Southampton

See also in the *Berkshire Encyclopedia of Sustainability* Energy Efficiency; Energy Industries–Overview of Renewables; Energy Industries–Wind; Investment, CleanTech; Water Use and Rights

Further Reading

Baker, A. Clive. (1991). *Tidal power.* London: Peter Pereginus.

Charlier, Roger H. (2007). Forty candles for the Rance River TPP tides provide renewable and sustainable power generation. *Renewable and Sustainable Energy Reviews*, *11*(9), 2032–2057.

Douglas, C. A.; Harrison, Gareth P.; & Chick, John P. (2008). Life cycle assessment of the Seagen marine current turbine. *Proceedings of the Institution of Mechanical Engineers, Part M: Journal of Engineering for the Maritime Environment*, *222*(1), 1–12.

Falnes, Johannes. (2007). A review of wave-energy extraction. *Marine Structures*, *20*(4), 185–201.

Fraenkel, Peter L. (2007). Marine current turbines: Pioneering the development of marine kinetic energy converters. *Proceedings of the Institution of Mechanical Engineers, Part A: Journal of Power and Energy*, *221*(2), 159–169.

Garrett, Chris, & Cummins, Patrick. (2008). Limits to tidal current power. *Renewable Energy*, *33*(11), 2485–2490.

International Ship and Offshore Structures Congress. (2006). Specialist Committee V.4: Ocean, wind and wave energy utilization. *Proceedings of the 16th International Ship and Offshore Structures Congress*, *2*, 165–211.

International Ship and Offshore Structures Congress. (2009). Specialist Committee V.4: Ocean, wind and wave energy utilization. *Proceedings of the 17th International Ship and Offshore Structures Congress*, *2*, 201–258.

Nicholls-Lee, Rachel F., & Turnock, Stephen R. (2008). Tidal energy extraction: Renewable, sustainable and predictable. *Science Progress*, *91*(1), 81–111.

United States Energy Information Administration (USEIA). (2009). Table 5: US average monthly bill by sector, census division, and state 2007. Retrieved January 14, 2010, from http://www.eia.doe.gov/cneaf/electricity/esr/table5.html

Energy Industries—Wind

In the early twenty-first century, wind power supplies a very small percentage of energy worldwide for electricity generation. But as a renewable source with no carbon emissions, there is global interest in increasing its role. The high cost of wind turbines, reaching demand centers, its relative unpredictability, and the variation in actual wind speed (not the theoretical capacity) are its primary disadvantages.

Approximately 50 percent of the electricity consumed in the United States in 2008 was produced using coal as fuel. According to the United States Energy Information Administration (USEIA), the independent statistical agency of the US Department of Energy, coal-fired generation accounted for 41 percent of world electricity supply in 2006, the latest year for which data are available (USEIA 2009e, 74). In light of the scientific evidence documented by the Intergovernmental Panel on Climate Change (2007) and other organizations, this is clearly not sustainable; based on data reported by the EIA, for every kilowatt-hour of electricity produced using coal, approximately 1 kilogram of harmful carbon dioxide is released into the atmosphere (USEIA 2009c, 106)

One sustainable alternative to coal is wind energy. According to the Energy Information Administration, the share of electricity generation in the United States from wind turbines was 1.3 percent in 2008 (USEIA 2009c, 11). In 2007, 3.7 percent of the European electricity demand was met by wind power (European Wind Energy Association 2009). Because there are no carbon emissions associated with wind energy, there is considerable support for increasing this share.

The US Department of Energy has indicated that it is feasible for wind energy to supply 20 percent of US electricity consumption by 2030 (US Department of Energy 2008). The European Union (EU) has set a binding target to provide 20 percent of its electricity supply from wind and other renewables (European Wind Energy Association 2009). The achievement of these targets is dependent on the cost of wind power, energy regulatory policies, transmission access, climate legislation, and the success of electricity system managers in integrating wind energy into their operations.

Cost of Wind Energy

The Energy Information Administration has reported that the capital costs of a new wind project can be expected to be more than double the capital costs of an equivalent-sized conventional power plant (USEIA 2009a, 93). Globally these capital costs are dominated by the price of turbines (European Wind Energy Association 2009). Despite the savings from not purchasing fuel (as would be the case for a conventional power plant), this capital-cost disadvantage has tended to discourage investment in wind energy. Compounding this drawback is the fact that variations in wind speeds mean that the annual wind-energy production level from a typical onshore wind turbine is only about 40 percent of its theoretical capacity (USEIA 2009a, 161).

Government intervention, as well as high fossil-fuel prices, has allowed wind in some countries to come close to achieving what is known as "grid parity" (Komor 2009). Grid parity occurs when electricity generated from a renewable source is cost competitive with electricity generated from more traditional sources such as coal and natural gas. An example of government intervention in the cost of wind power is the US production tax credit (PTC), which provides developers of renewable energy projects a ten-year credit of approximately $20 per megawatt-hour (MWh) generated, which is about 20–30 percent of the generation costs by wind (US Department of Energy 2008, 28). (A megawatt-hour is a measure of energy equal to one thousand kilowatt-hours; the average US home consumes 0.94 MWh in a month [calculation based on data from USEIA 2009f]. For more on this topic please refer to the sidebar "What's the difference between a gigawatt and a gigawatt-hour?" on pages 4–5.) Originally passed as part of the Energy Policy Act of 1992, the PTC has been extended repeatedly and also has been permitted to lapse three distinct times since its inception. When the US Congress has allowed this credit to expire in the past, the development of wind resources has significantly decreased, demonstrating the importance of an incentive to the industry.

Subsidies are also an important driver of wind energy development in Europe. In Germany, the Electricity Feed Law (StrEG),adopted in 1991, obligates public utilities to purchase renewably generated power from wind, solar, hydro, biomass, and landfill gas sources on a yearly fixed-rate basis, based on the utilities' average revenue per kilowatt-hour (kWh). Compensation to wind producers was set at 90 percent of the average retail electricity rate (Runci 2005). This is substantially higher than the wholesale price received by traditional generators. This renewable energy payment—a "feed-in tariff"—has since been modified, but at this point, twenty-one European countries have introduced some form of feed-in tariff (Crystall 2009).

Wind Energy Intermittency

Under currently technology, it is not economically viable to store large quantities of electricity. Moreover, the stability of an electricity grid requires that that the amount of power generation in a given "control area" area matches exactly, on a near-instantaneous basis, the system load, net of losses, and electricity flows with other control areas. Unfortunately the production of electricity using wind turbines exhibits largely uncontrollable variability over the course of a day (the upward deployment of wind energy is not possible). This is illustrated in figure 1, which depicts wind energy production levels in the ERCOT power grid in Texas by fifteen-minute intervals over the month of September 2009. (ERCOT accounts for about 85 percent of the electricity load in Texas.) Inspection of data for other control areas such as western and eastern Denmark, two of the world's most wind-intensive power systems, reveals a similar level of variability in the hourly production over the same time period (Energinet.dk 2009). This variability in production needs to be compensated for by dispatching conventional energy and/or importing power from other regions. Operators of power grids attempt to adapt to this variability in production by forecasting wind energy production levels on a day-ahead basis using meteorological data.

Unfortunately it is difficult to accurately forecast wind energy production levels on a day-ahead basis. A recent study (Forbes, Stampini, and Zampelli 2010) indicates that the ERCOT's mean day-ahead wind-energy forecasting error rate was more than 50 percent of the average wind energy production from the period 15 June 2009 through 30 November 2009. The ability to store electricity in large quantities that could be used when the actual level of wind energy is less than forecasted is one possible solution to this challenge. Fortunately in the United States there are several Department of Energy–funded projects underway that are designed to advance this solution (US Department of Energy 2007).

Transmission Access

In the United States, wind resources are typically most robust in remote interior locations such as the Great Plains and the Rocky Mountains, while most of the population lives along the East and West coasts (American Wind Energy Association 2009). Because of this mismatch between resources and demand centers, the US Department of Energy has identified electricity transmission as the largest obstacle to the development of wind energy resources. Highlighting the magnitude of the challenge, when the cost of electricity transmission is ignored, the United States is believed to have more than 8,000 gigawatts (GW) of wind resources that the industry estimates can be developed at a cost of less than or equal to approximately $80 per MWh (US Department of Energy 2008, 8). To put this number in perspective, the United States currently has about 1,000 GW of conventional electricity-generating capacity. When transmission costs are factored in, only about 600 GW of wind energy resources can be developed at a cost of less than or equal to a delivered wholesale cost of approximately $100 per MWh (US Deparment of Energy 2008, 9). Even at a price of $100 per MWh, the state of the transmission grid in 2009 is inadequate to support the development of significant amounts of

Figure 1. September 2009 Quarter-Hour Wind Energy Production Levels in the Electric Reliability Council of Texas (ERCOT) Power Grid

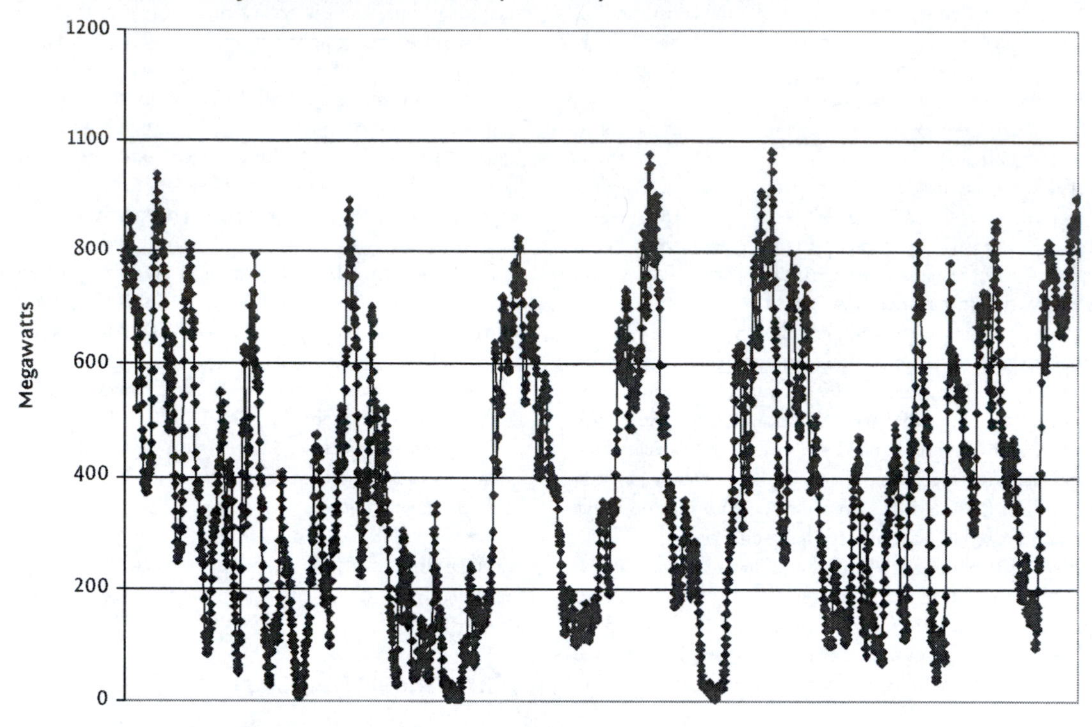

Note: The data are reported in fifteen-minute intervals.

Source: Forbes, Stampini, and Zampelli (2010).

As an energy source, the wind can be variable, uncontrollable, and unpredictable, which influences its reliability for electricity generation.

wind resources. The infrastructure in the United States is antiquated, which causes capacity issues and interconnection backlog. Moreover, there is an insufficient level of transmission capacity from the regions with the best wind resources (Komor 2009). The transmission issue must be addressed in order for wind to become a more integral piece of energy supply in the United States.

According to the European Wind Energy Association (EWEA), transmission grid improvements and fair interconnection rules also must be instituted in Europe in order to achieve the targeted amounts of wind energy penetration. It is estimated that in many areas of Europe it could take ten to twelve years to construct a new transmisssion line due to lengthy planning and permission processes (EWEA 2009). Countries such as Denmark, however, have been able to achieve high levels of reliance on wind energy due in part to their concerted effort to link their tranmission system with other countries (Komor 2009).

Energy Regulatory Policies

Regulatory policies have been and will continue to be a key driver of renewable energy development (Wiser and Barbose 2008). The wind industry has gained particular momentum since the 1990s as regulatory mandates have been passed in Europe and the United States. In some countries, including the United States, mandatory goals have been established requiring utilities to include a certain percentage of renewable energy in their supply portfolio; this is commonly known as renewable portfolio standard (RPS). The percentages typically increase on an annual basis, and the accompanying regulations usually include a penalty if the goals are not met. As of July 2009, twenty-nine states and the District of Columbia had RPS policies in place (North Carolina Solar et al. 2009). In addition, in June 2009 the US House of Representatives passed H.R. 2454, the American Clean Energy and Security Act of 2009 (ACESA), which requires retail energy suppliers to have at least 20 percent of their portfolio originate from renewable sources by 2020. Denmark, the Netherlands, Austria, Italy, Belgium, and the United Kingdom, all of which have significantly higher levels of wind energy penetration, have also adopted or proposed RPS programs (Geller 2003). For example, approximately 20 percent of electricity consumption in Denmark during 2008 was generated using wind energy (Energinet.dk 2009). This achievement is a clear example of how government intervention can increase the use of wind power: Denmark has invested more in wind energy than any other European nation, namely through subsidies and a strong financial commitment to research and development.

Although it is difficult to say definitively that RPS policies are the primary reason renewable energy projects are built in certain areas, it is clear that countries and US states that have RPS policies in place have seen an increase in renewable development (Wiser and Barbose 2008). Strong renewable policy (such as an RPS) signals the market and spurs capital investment. Projects are typically built by independent power producers who sell the power and any associate environmental attributes associated with their projects to utilities with an RPS obligation. Wind power has seen the largest growth as RPS has become more popular (Geller 2003). Specifically from 1998 to 2007 in the United States, 93 percent of all non-hydro renewable energy capacity that has come online in RPS states has come from wind power (Wiser and Barbose 2008). Although the US wind industry is enjoying strong policy support as of 2009, it continues to urge the federal government to adopt a national RPS, thereby providing a wide signal for capital investment (American Wind Energy Association 2009). The EU has set such a binding renewable target, which is expected to spur continued investment throughout Europe (EWEA 2009).

In some US states, renewable energy credits (RECs) can be used by utilities to comply with their RPS obligations. An REC is a separately tradable certificate of proof that one megawatt-hour of electricity was generated from a renewable energy source. When an RPS program allows RECs to be used for compliance, sellers of wind energy have access to two distinct revenue streams, one for the REC and the other for the energy itself. Although REC prices vary from state to state, the revenue from REC sales can be considerable. Take for example, the price of a wind REC in Pennsylvania, which was worth between $10 and $13 in 2008, while the average day-ahead wholesale energy price in the same region was approximately $70. In addition there are several optional green programs whereby RECS are available to end-use customers for a premium.

The Future

The technical potential of wind energy is substantial; thousands of gigawatts of wind energy capacity have the potential to be developed. Because of favorable government policies, wind energy is currently one of the fastest-growing supply sources of electricity throughout the world. The Global Wind Energy Conference (GWEC) forecasts a 119 percent increase in wind energy capacity from 2009, reaching 332 GW of installed capacity by 2013 (GWEC 2008, 15). The economics of wind energy are challenging, however, and thus in order for wind energy to be more than a niche source of electricity, there needs to be a cap on carbon emissions, which will make wind energy more economically attractive. In addition, substantial investments in transmission are needed to move the wind energy to demand centers, and research and development in energy storage needs to be accelerated to address the issue of intermittency.

Kevin F. FORBES
The Catholic University of America

Adrian DiCianno NEWALL
Energy Consultant

See also in the *Berkshire Encyclopedia of Sustainability* Cap-and-Trade Legislation; Climate Change Disclosure; Energy Efficiency; Energy Industries—Overview of Renewables; Energy Industries—Coal; Energy Industries—Wave and Tidal; Investment, CleanTech

Further Reading

American Wind Energy Association (AWEA). (2009). *Windpower outlook 2009.* Retrieved October 8, 2009, from http://www.awea.org/pubs/documents/Outlook_2009.pdf

Bertoldi, Paolo; Atanasiu, Bogodan; European Joint Research Commission; & Institute for Environment and Sustainability. (2007). Electricity consumption and efficiency trends in the enlarged European Union: Status report 2006. Retrieved January 14, 2010, from http://re.jrc.ec.europa.eu/energyefficiency/pdf/EnEff%20Report%202006.pdf

Crystall, Ben. (2009, September 15). Better world: Generate a feed-in frenzy. Retrieved Ferburary 5, 2010, from http://www.newscientist.com/article/mg20327251.900-better-world-generate-a-feedin-frenzy.html

Energinet.dk. (2009). *Annual report 2008 of Energinet.dk.* Retrieved October 8, 2009, from http://www.energinet.dk/NR/rdonlyres/876A786B-4646–4FB0–8C0E-D367D6274F90/0/Annual_Report_2008.pdf

European Wind Energy Association (EWEA). (2009). *Wind energy—The facts: Analysis of Wind Energy in the EU-25.* Retrieved February 5, 2010, from http://windfacts.eu/

Forbes, Kevin; Stampini, Marco; & Zampelli, Ernest M. (2010). *Do higher wind power penetration levels pose a challenge to electric power security?: Evidence from the ERCOT Power Grid in Texas.* Unpublished manuscript.

Geller, Howard. (2003). *Energy revolution: Policies for a sustainable future.* Washington, DC: Island Press.

Intergovernmental Panel on Climate Change. (2007). *Climate change 2007: The physical science basis: Working Group I contribution to the Fourth Assessment Report of the Intergovernmental Panel on Climate Change, 2007.* New York: Cambridge University Press.

Global Wind Energy Conference (GWEC). (2008). *Global wind: 2008 report.* Retrieved February 9, 2010, from http://www.gwec.net/fileadmin/documents/Global%20Wind%202008%20Report.pdf

Komor, Paul. (2009). *Wind and solar electricity: Challenges and opportunities.* Arlington, VA: Pew Center on Global Climate Change. Retrieved February 5, 2010, from http://www.pewclimate.org/docUploads/wind-solar-electricity-report.pdf

North Carolina Solar Center; Interstate Renewable Energy Council; United States Department of Energy, Energy Efficiency and Renewable Energy; & National Renewable Energy Laboratory. (2009). *Database of state incentives for renewables & efficiency: Rules, regulations & policies for renewable energy.* Retrieved December 8, 2009, from http://www.dsireusa.org/summarytables/rrpre.cfm

Runci, Paul J. (2005). *Renewable energy policy in Germany: An overview and assessment* (Pacific Northwest National Laboratory technical report PNWD-3526). Retrieved February 9, 2010, from http://www.globalchange.umd.edu/data/publications/PNWD-3526.pdf

United States Department of Energy (DOE). (2007). Basic research needs for electrical energy storage. Retrieved October 8, 2009, from http://www.sc.doe.gov/bes/reports/abstracts.html#EES

United States Department of Energy (DOE). (2008). 20% wind energy by 2030: Increasing wind energy's contribution to US electricity supply. Retrieved October 8, 2009, from http://www1.eere.energy.gov/windandhydro/pdfs/41869.pdf

United States Energy Information Administration (USEIA). (1999). Natural gas 1998: Issues and trends. Retrieved January 19, 2010, from http://www.eia.doe.gov/pub/oil_gas/natural_gas/analysis_publications/natural_gas_1998_issues_trends/pdf/chapter2.pdf

United States Energy Information Administration (USEIA). (2008). Federal financial interventions and subsidies in energy markets 2007. Retrieved October 8, 2009, from http://www.eia.doe.gov/oiaf/servicerpt/subsidy2/index.html

United States Energy Information Administration (USEIA). (2009a). Assumptions to the annual energy outlook 2009. Retrieved October 8, 2009, from http://www.eia.doe.gov/oiaf/aeo/assumption/index.html

United States Energy Information Administration (USEIA). (2009b). Energy market and economic impacts of H.R. 2454, the American Clean Energy and Security Act of 2009. Retrieved October 8, 2009, from http://www.eia.doe.gov/oiaf/service_rpts.htm

United States Energy Information Administration (USEIA). (2009c). Electric power annual. Retrieved February 5, 2010, from http://www.eia.doe.gov/cneaf/electricity/epa/epa.pdf

United States Energy Information Administration (USEIA). (2009d). Electric power monthly. Retrieved October 8, 2009, from http://www.eia.doe.gov/cneaf/electricity/epm/table1_1.html

United States Energy Information Administration (USEIA). (2009e). International Energy Outlook 2009. Retrieved January 27, 2010, from http://www.eia.doe.gov/oiaf/ieo/electricity.html

United States Energy Information Administration (USEIA). (2009f). Table 5: US average monthly bill by sector, census division, and state 2007. Retrieved January 14, 2010, from http://www.eia.doe.gov/cneaf/electricity/esr/table5.html

Wiser, Ryan, & Barbose, Galen. (2008). *Renewable portfolio standards in the United States: A status report with data through 2007.* Retrieved October 8, 2009, from http://eetd.lbl.gov/ea/ems/reports/lbnl-154e.pdf

Energy Labeling

Recent awareness on process and product sustainability has brought about the diffusion of a wide range of tools to measure, evaluate, and compare energy performance. Energy labeling represents one of the most important systems in providing benchmarks for both sustainability performance and guidance in purchases models.

The increasing concern about business impacts on the environment and society has resulted in growing attention toward sustainable production and consumption models. Currently a wide range of tools providing significant benchmarks for sustainability product performance as well as guidance for consumers is available.

Energy labels are informative labels applied to manufactured products indicating data relative to energy performance, generally in terms of consumption, efficiency, cost, and so on. Consumers thus are provided with the necessary information for making more-informed choices.

Currently, three categories of energy labels are used in most countries: endorsement, comparative, and information only. Endorsement labels essentially offer a "seal of approval" that a product meets certain prespecified criteria. They are generally based on a "yes/no" procedure and offer little additional information. One example of an endorsement label for energy efficiency is the Energy Star label that is provided by the US Environmental Protection Agency (EPA).

Comparative labels are divided into two subcategories: one involves a categorical ranking system, and the other uses a continuous scale or bar graph to show relative energy use. The category labels use a ranking system that tells consumers how energy efficient a model is compared to others. The main emphasis is on establishing clear categories so that the consumer can easily understand, by looking at a single label, how an energy-efficient product compares relative to others in the market. The European energy label is an example of a category label. (See figure 1 on page 90.)

The other category of comparative label—continuous-scale labels—provide comparative information that enables consumers to make informed choices about products; however, they do not concern specific categories. The Canadian energy guide is an example of the continuous-scale label. (See figure 2 on page 90.)

Information-only labels provide data on the technical performance of the labeled product and offer no simple way (such as a ranking system) to compare energy performance between products. These types of labels are generally not consumer friendly because they contain only technical information. (See figure 3 on page 91.)

It is important to keep a consistent label style and format across product types; this makes it easier for consumers to understand individual types of labels to evaluate different products. Selecting a label to use is not always easy and usually depends on local consumer knowledge and attitudes. The endorsement label is quite effective, at least with consumers that are attentive to environmental issues. Categorical comparison labels provide more information about energy use and, if well designed and implemented, can provide a consistent basis that buyers can focus on when evaluating energy efficiency from one purchase to another. Continuous-scale labels can transmit more detailed information on relative energy use, but research has shown that this label format may be difficult for consumers to understand. Information-only labels are generally more effective for the most educated and economically and/or environmentally concerned consumers.

Figure 1. European Union Color Energy Label

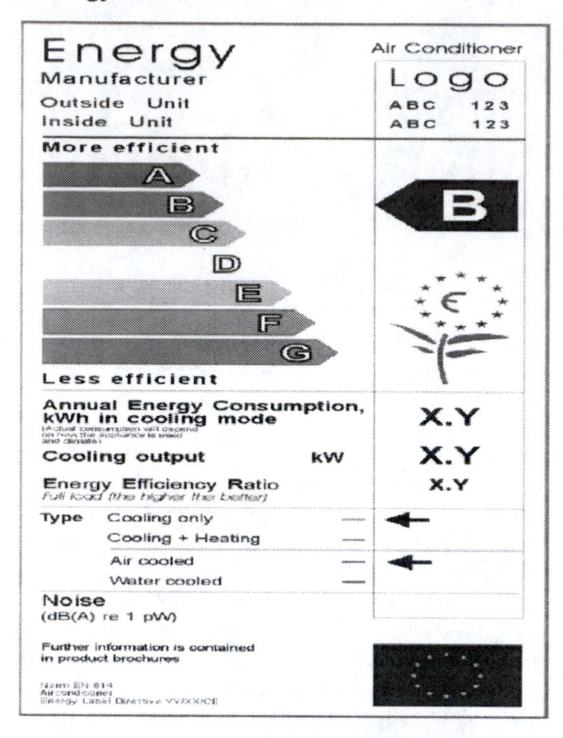

Source: ENEA (2004).

In this European Union label, the energy efficiency of an appliance, in this case an air conditioner, is rated in terms of a set of energy efficiency classes from A to G on the label, A (green) being the most energy efficient, G (red) the least efficient. The label also gives customers other useful information when choosing between various models.

Figure 2. Continuous-Scale Label (Canadian Energy Label)

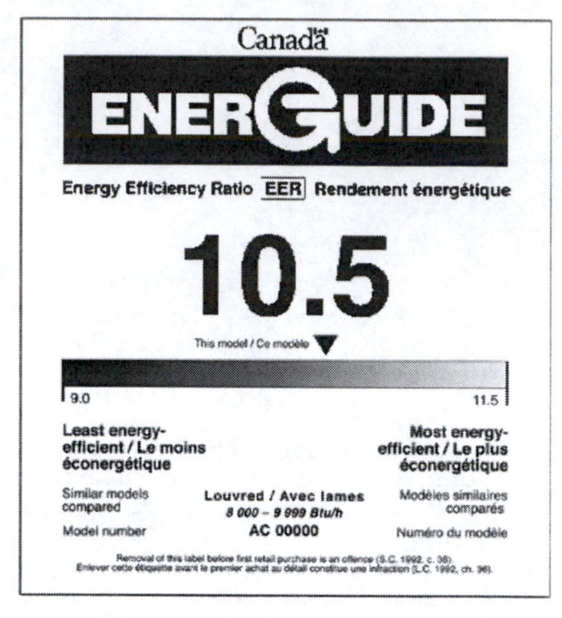

Source: Natural Resources Canada (2010).

Canada's black and white EnerGuide Label shows the average annual energy consumption of the appliance in kilowatt hours (kWh), energy efficiency of the appliance relative to similar models, annual energy consumption range for models of this type and size, type and size of the model, and model number.

Energy Labeling Overview

National energy labeling programs were developed at the beginning of the 1970s with the goal of controlling the confused overflow of "self-certified" private energy brands as well as for facilitating informed consumer choices. One of the first forms of these energy-efficiency labeling programs—obligatory for domestic-use refrigerators, freezers, and air-conditioners—dates to Canada's 1978 EnerGuide program, which was run by the government agency National Resources Canada. Subsequently a similar program was developed in the United States called Energy Guide, jointly managed by the Department of Energy (DOE) and the Environmental Protection Agency (EPA). In Europe, the Directive 79/530/CEE established that household energy consumption of electrical appliances had to be indicated. Unfortunately, this never transposed because this directive was never implemented in all European countries. In Germany in 1986, the Blauer Engel plan considered marking boilers, while the Dutch program Milieuker began labeling light bulbs, computers, and televisions. The almost simultaneous 1986 diffusion of the Star Rating Scheme on the Australian continent was the outcome of the mandatory energy labeling of refrigerators, while in Canada the Environmental Choice Program (1988), an environmental excellence label, was envisaged for both domestic appliances as well as office equipment.

In the 1990s, following the success of previous initiatives, further energy labeling programs became widespread throughout the rest of the world. The European Union (EU) Directive 92/75/CEE relative to a mandatory energy labeling program provided energy consumption indications and supplementary information on how domestic appliances work. At the same time in the United States, in the light of previous experience in the household domestic appliances sector, an energy efficiency label—one of the most important on an international level—was applied to computers and computer

Figure 3. Information-Only Energy Label (US Energy Guide)

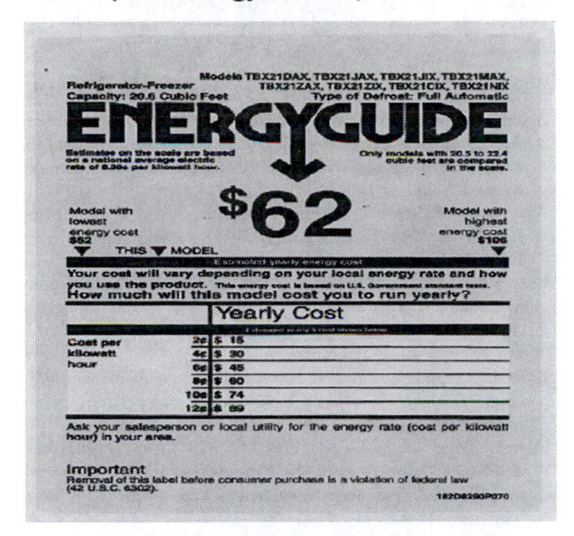

Source: Federal Trade Commission (2010).

The black and white US Energy Guide shows average energy consumption of the appliance in kilowatt hours (kWh); energy efficiency of the appliance; and the relative cost and energy consumption range for different models.

monitors (Energy Star Program) as well as an environmental excellence label (Green Seal) for light bulbs.

Energy Star is an international voluntary labeling mechanism introduced by the EPA in 1992. The agency's main aim was to promote the most energy-efficient products available on the market in order to facilitate energy saving as well as reduce climate-altering greenhouse gas emissions. The Energy Star label is currently the most widespread internationally and applies to over forty different types of products, including household electrical appliances (washing machines, refrigerators, dishwashers), office equipment (faxes, printers, scanners, computers), lighting products, home electronics, and more. The program merits particular attention as it enables the certification of buildings from an energy performance point of view (i.e., buildings with low energy consumption). During the late 1990s, the GEEA label was created in Switzerland by the Group for Energy Efficient Appliances (GEEA), a forum of government agencies from various countries including Denmark, Holland, and Sweden. The label—applied to computer appliances as well as to televisions, video recorders, DVD players, satellite receivers, audio systems, and more—has become widely used in the GEEA countries. The efficiency criteria—to which the appliance must conform—are far more restrictive when compared to the Energy Star label. In Sweden, following several pioneering experiences within the certification sector, in particular regarding computer appliance safety, the Tjänstemännens Central Organization (TCO, or the Swedish Confederation of Professional Employees) label TCO'95 for cathode-ray tube monitors and computer keyboards was adopted. After several years, the program became more widespread—through the TCO'99 label—and was applied to liquid crystal monitors, printers, and laptop computers. The conforming criteria for label use include ergonomic aspects, radiation emissions reduction, low energy consumption, and appliance eco-efficiency.

Meanwhile in South America, the Programa Brasileiro de Etiquetagem, another type of energy labeling program, was set up and voluntarily applied to refrigerators and air conditioning by manufacturers. At the same time following the growth of the information and communication technology (ICT) sector, Asia implemented the Japanese environmental excellence label Eco Mark for photocopiers (1999), computers (2000), and printers (2001). The US government, one of the most attentive toward the competitive strategies being applied to its own market, adopted the International Energy Star program, followed closely by others including Japan (1995), Australia (1999), New Zealand (1999), the EU (2000), Taiwan (2000), and Canada (2001). The choice was based on bilateral agreements made with these countries. Subsequently, another environmental excellence label, the Good Environmental Choice, was applied to the computer sector in Australia.

The Swedish TCO had proposed the TCO'03 display label to be used specifically for computer monitors. TCO'03 not only represented an innovative generation of labeling programs, based on more restricting environmental and energy criteria than its predecessors, but it also envisaged the labeling of monitors that had superior performance qualities, such as color and image reproduction, low radiation emissions, safety, and energy efficiency. This could reduce the environmental impact during the product's life cycle, as evaluated through a life cycle assessment (LCA).

During the last decade, an agreement was signed between the US government and the European Union to coordinate the labeling programs relative to efficient energy use in office equipment. This resulted in the EU Directive of 8 April 2003 that definitively established the Energy Star program, which represents a decisive step toward harmonizing not only all the different types of energy consumption certification programs but also the diffusion of more energy efficient products. The European Union issued the Directive 2010/30/EU, which extends a new system of energy labeling to products that consume energy for commercial and industrial use, for instance television sets, decoders, fittings, refrigerators, and industrial machinery.

Implications

Energy labeling programs can realistically assume a driving force role in the transition process toward eco-sustainability only if their current critical elements are eliminated.

Without doubt, the main critical element is consumer disorientation that derives from the articulated range of energy labeling systems on the market. When the consumer is not always able to fully understand the information on the label, specific technical competence often is required. A further critical aspect involves the criteria adopted in the different programs, which at times is far too complex and on other occasions, too vague. The combination of these factors strongly limits the diffusion of energy-efficient products. Moreover, the price factor, which is generally higher than similar products, also represents a relevant barrier. In fact, the consumer is not always willing to pay a higher price for the purchase of energy-efficient goods.

No doubt, more widespread diffusion and development of energy labeling programs that are harmonized on a global level is desirable. Innovative solutions depend on consumer choices that are still based on price, without even considering socio-environmental advantages. The transition to a more eco-sustainable lifestyle and consumption model (both local and global) implies establishing agreements on various levels for the diffusion of standard programs, energy efficiency logos, and specific types of products, all capable of correctly informing the consumer on the economic, environmental, and social advantages of the product.

Ornella MALANDRINO
Salerno University

See also in the *Berkshire Encyclopedia of Sustainability* Building Rating Systems, Green; Ecolabels; Energy Efficiency Measurement; Organic and Consumer Labels

FURTHER READING

Agenzia nazionale per le nuove tecnologie (ENEA) [Italian National Agency for New Technologies]. (2004). *L'etichetta energetica* [The energy label] (Opuscolo no. 24 della Collana Sviluppo Sostenibile). Rome: ENEA.

Agenzia nazionale per le nuove tecnologie (ENEA) [Italian National Agency for New Technologies]. (2011). Homepage. Retrieved February 3, 2011, from http://www.enea.it/it

Energy Star. (2011). Homepage. Retrieved February 3, 2011, from http://www.energystar.gov

EU Energy Star. (2011). Homepage. Retrieved February 3, 2011, from http://www.eu-energystar.org

Europa: Il portale dell'Unione Europea [Gateway to the European Union]. (2011). Homepage. Retrieved February 3, 2011, from http://europa.eu/index_it.htm

Federal Trade Commission (FTC). (2010). Energy guidance: Appliance shopping with the Energy Guide label. Retrieved June 29, 2011, from http://www.ftc.gov/bcp/edu/pubs/consumer/homes/rea14.shtm

Harrington, Lloyd, & Damnics, Melissa. (2004). Energy labeling and standards programs throughout the world. Canberra, Australia: National Appliance and Equipment Energy Efficiency Committee (NAEEEC).

Mahlia, T. M. Indra. (2004). Methodology for predicting market transformation due to implementation of energy efficiency standards and labels. *Energy Conversion and Management, 45*(11), 1785–1793.

Mahlia, T. M. Indra, & Saidur, Rahman. (2010). A review on test procedure, energy efficiency standards and energy labels for room air conditioners and refrigerator-freezers. *Renewable and Sustainable Energy Reviews, 14*(7), 1888–1900.

Natural Resources Canada. (2010). Household appliances. Retrieved June 29, 2011, from http://oee.nrcan.gc.ca/residential/personal/appliances/energuide.cfm?attr=4#household

Proto, Maria; Malandrino, Ornella; & Supino, Stefania. (2007). Ecolabels: A sustainability performance in benchmarking? *Management of Environmental Quality: An International Journal, 18*(6), 669–683.

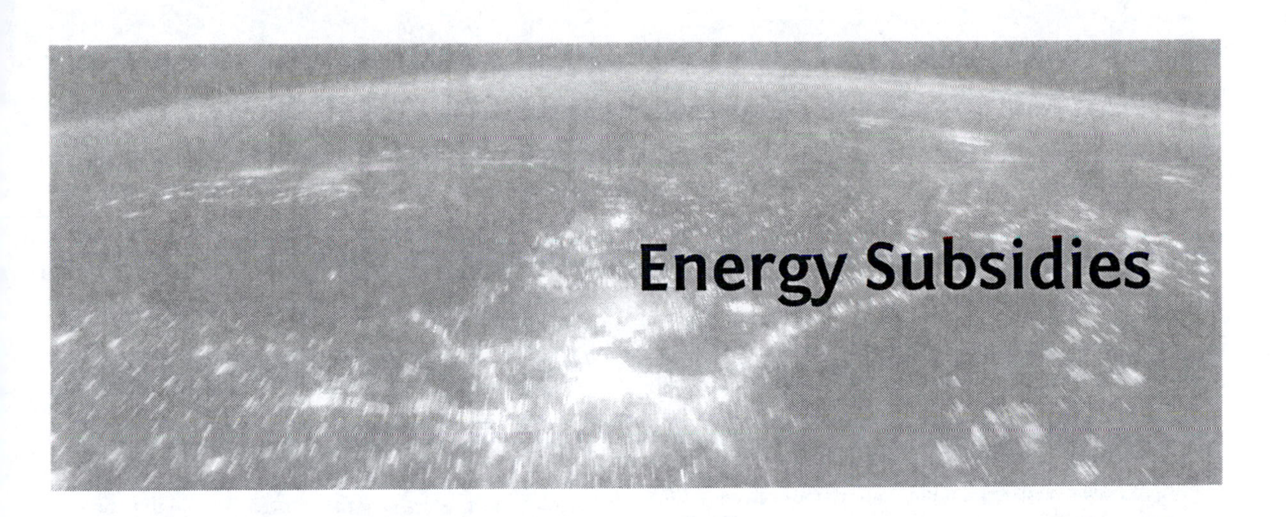

Energy Subsidies

Government-funded energy subsidies are used around the world to help expand access to energy sources and increase energy production. Almost all energy sources have received subsidies at some point. Energy subsidies play, and will continue to play, a major role in global energy policies for both sustainable and traditional resources.

Energy subsidies have long been used to help promote, support, and expand access to energy sources. The producers and consumers of virtually every type of energy source, from fossil fuels to renewable and sustainable resources, have received subsidies. This is true for more traditional sources, such as hydropower, coal, oil, and gas, as well as for newer or emerging sources, such as nuclear, biomass, wind, and solar energy.

Subsidy programs have been a part of most government energy policies used to expand access to energy, and such subsidies are likely to play a significant role in expanding access to sustainable energy sources as well. These subsidies vary depending on location, motivation, and political ideology.

Types of Energy Subsidies

There are four primary types of energy subsidies. These government-funded programs can take many forms, but typically take one of the following forms: (1) direct spending, (2) tax reduction, (3) support for research and development (R&D), and (4) government-run programs facilitating access.

Direct Spending

Direct spending programs are programs through which the government provides payments directly to either consumers or producers of energy. These payments can be payments for the production of a certain amount of the supported resource or payments directly to consumers.

One example of a direct subsidy is a targeted program such as the Low Income Home Energy Assistance Program (LIHEAP). LIHEAP is a federal program in the United States designed to help low-income households with their energy bills, primarily by supporting immediate energy needs.

Direct subsidies could also be part of regulatory programs or taxes designed to improve the environment or combat climate change. As some recent legislative proposals have suggested, proceeds from a cap-and-trade program or a carbon tax could be used to fund subsidies for renewable and sustainable energy projects.

Tax Reductions

Tax reduction programs can take the forms of tax deductions (reducing taxable income upon which taxes are calculated) and tax credits (reducing an overall tax obligation). The reductions can be linked to energy production or investment in energy infrastructure.

Production Tax Credits

One common energy subsidy takes the form of a production tax credit (PTC). This credit is calculated by multiplying the credit amount (e.g., 2.1 cents per kilowatt hour) by the amount of power generated (e.g., 80 kilowatt hours). The PTC is paid each year for the duration of the specified credit period, which varies by law and often by the type of energy produced. A PTC usually has restrictions on who can receive the generated power, such as a requirement that the generated power be sold to an unaffiliated purchaser. Finally, because the PTC is used to support developing or (relatively) costly energy sources, most PTCs will have a price cap that gradually phases out the credit as the price of the generated power increases.

Investment Tax Credit

Another often used renewable and sustainable energy subsidy is the investment tax credit (ITC). Rather than basing the credit on energy production, an ITC is based on the cost of the renewable or sustainable energy facility or the property costs for such a facility. Eligible facilities are determined by statute, and often include projects with high start-up costs, such as geothermal technologies, solar projects, and nonutility-scale wind projects. ITCs provide a credit for a certain percentage of the project costs; the credit is typically vested over a certain time frame (e.g., an ITC of 30 percent of the project could be captured by the developer at 20 percent of the credit per year).

Consumer Energy Credits and Tax-Free Grants

Consumer-side grants and credits can also be used to promote sustainable energy practices. These subsidies can take the form of tax credits, which reduce the amount of tax owed dollar for dollar, or tax deductions, which provide for a certain percentage reduction of the tax owed. Examples of consumer-side tax credits or deductions include home energy efficiency credits (e.g., replacing windows), residential renewable energy credits (e.g., solar water heaters), and credits for renewable or alternative fuel vehicles. Finally, some governments use tax-free grants to promote energy efficiency and sustainable energy practices. Canada, for example, provides up to $5,000 in tax-free grants for residents who undergo an energy efficiency audit before renovating.

Research and Development

Research and development (R&D) subsidies are often used to increase energy supplies and to improve energy production efficiencies and technologies. Such subsidies are not expenditures that necessarily impact energy production or prices, but when the subsidies lead to useful technologies and processes they can impact future prices and rates of production.

R&D subsidies often take the form of government-sponsored grants that are used to help reduce or otherwise offset the initial risks related to developing or installing new technologies. In addition, R&D subsidies can be used to fund test projects for promising, but unproven, technologies or sites. Such grants are often part of a public–private partnership, where the governmental subsidy is matched with funds from other interested parties. As an example, a recent program funded through the US Department of Energy offered $338 million in government grant money for geothermal research and development, and private and other nonfederal sources provided additional funds exceeding that amount.

Government-Run Programs Facilitating Access

Government programs increasing access to energy are often programs targeted at specific regions. Such subsidy programs may provide government-funded programs that bring large amounts of electricity to market in the targeted region. These subsidy programs may also indirectly subsidize portions of the electricity industry through loans and loan guarantees to facilitate the construction of infrastructure necessary to make energy accessible.

Virtually all governments are providing subsidies to increase access to and the availability of energy. A good example of an early program is the success of the United States' Rural Electrification Act (REA), which provided the long-term financing and technical expertise needed to expand the availability of electricity to rural customers. In 1963, President John F. Kennedy explained that, since the passage of the REA in 1936, more than nine hundred cooperative rural electrification systems were built with the assistance offered by government-subsidized financing.

The REA's financial undertaking and related risk were enormous, thus necessitating subsidization. The program provided more than $5 billion to approximately 1,000 borrowers, facilitating construction of more than 1.5 million miles of power lines that, in the 1960s, served 20 million American people. In the end, the investment was remarkably sound. As of 1963, there was only one reported delinquent payment of those approximately 1,000 borrowers. The total expected losses on the $5 billion of financial assistance provided were less than $50,000. This low level of default is particularly striking in today's global financial environment.

Few investors were willing to invest in the rural electrification project without federal subsidization in the form of financing; the success of the project, however, was overwhelming. As an example, in 1963, North Dakota–based REA-funded cooperatives served an average of about one metered farm per mile of line, compared to the average urban-area utility system of thirty-three electric meters per mile of line, thus serving an amazing 97 percent of the state's population.

Regardless of the financing concerns, the subsidies were justified because it was believed that the REA raised the standard of living, strengthened the US economy, and improved national security by providing the power necessary to increase industrial activity. President Kennedy

(1964) explained the effects of the government subsidy to electrify rural America: "What was 30 years ago a life of affluence, in a sense today is a life of poverty."

The current need for energy and infrastructure subsidies in emerging markets is similarly essential to the need found in prior years in the United States. Subsidies in these emerging markets—including China, India, and major parts of Africa—are expected to be substantial, as access to basic energy sources becomes a reality for millions of people currently lacking significant, if any, access to electricity. These subsidies could potentially promote sustainable energy sources, but that is by no means a foregone conclusion. In the United States, the REA subsidies promoted increased access to energy of all sources, but especially fossil fuel sources. In fact, today more than 90 percent of North Dakota's power still comes from coal-fired electricity generators.

That same infrastructure, however, can be used to move renewable-sourced electricity, if desired. As such, all over the United States the infrastructure subsidized to promote access electricity could be used to help promote access to renewable and sustainable energy sources.

The Role of Subsidies in Energy Policy

Although the type and extensiveness of government intervention varies widely, most governments use energy subsidies as some portion of their overall energy policy. These subsidies were traditionally used to support energy production and development, increase access to energy, and improve economic output. Many developed countries still provide subsidies for coal and natural gas extraction, and petroleum consumption is still subsidized in some of the major oil-exporting nations.

Today's subsidies often target the same goals, but the use of subsidies has evolved to include environmental goals and provide incentives for sustainable energy development. In pursuit of sustainable energy markets, many governments use subsidies and mandates to promote the use of sustainable energy sources. These subsidies have led to criticisms that renewable-sourced energy is being unfairly subsidized and promoted, even though more traditional forms of energy are less expensive (Ralls 2006, 452). Such criticisms ignore the fact that many traditional energy sources have costs that are not fully internalized, thus making them appear cheap, when, in fact, such sources are quite costly. That is, some costs of traditional sources, such as air pollution or carbon dioxide emissions that could cause global warming, are not part of the cost of consumption. These costs are often not added to—or internalized—the overall costs by consumers of fossil fuels.

Many who complain about renewable or sustainable energy subsidies argue that a "freer" market would provide more efficient markets and facilitate proper incentives for renewable and sustainable sources that are viable (Boaz 2005, 446). Removing renewable incentives in the name of free markets, however, means that all incentives related to energy sources should be eliminated. Only then can one really appreciate what "the market" actually wants. Arguing to reduce or eliminate subsidies for renewable energy sources is essentially a market-based argument cloaked in a source-based argument (e.g., a preference for fossil fuels over renewable sources). A true free-market argument recognizes that subsidies for traditional fuels are rampant as well, and would ask for a repeal of all such subsidies.

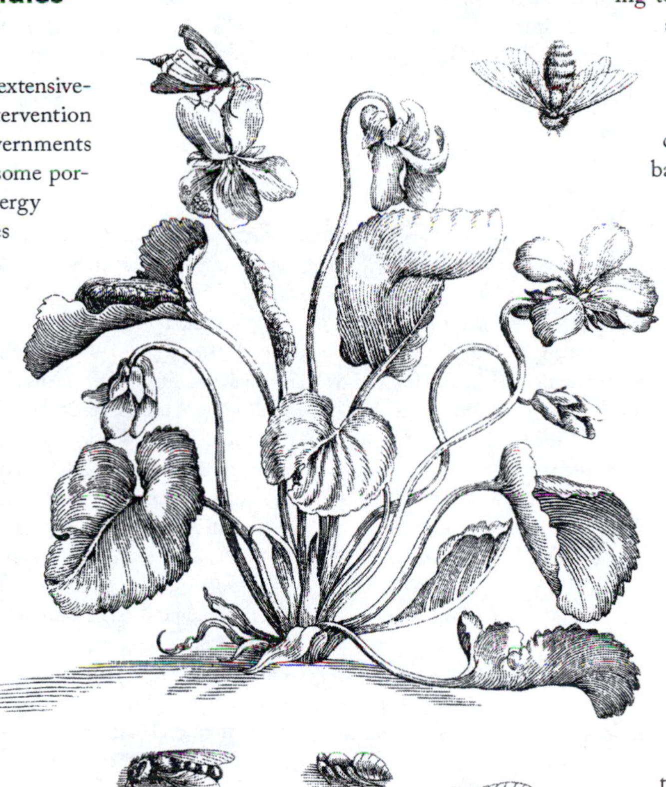

The incentives for traditional, nonrenewable fuel energy still outweigh renewable industry incentives in many places. Some argue that the idea that subsidies for conventional energy far outweigh those for renewable energy is somewhat misleading. That is, they argue that the amount of subsidies received per megawatt hour to support fossil fuels is significantly less than the subsidies received to support renewable energy (Lieberman 2010, 3). These subsidies can indicate different things, however, depending upon how one looks at the issue.

Most calculations of subsidies per source do not consider the full range of energy subsidies and their impact on markets, for example, ignoring the negative externalities flowing from some of the fossil fuel energy sources receiving subsidies. More specifically, using a per-unit comparison—considering a strict per-megawatt-hour amount of subsidies—does not look at the total market impact. As an example, a government loan to a small car company might amount to $5,000 per auto produced, while a similar loan to a major automaker might amount to only $2,000 per auto produced. The loan to the small automaker could be millions of dollars, while the loan to the large automaker would be a bailout worth billions of dollars. Thus, the per-auto number would not accurately reflect the actual market impact of the subsidies.

Recent US energy subsidies provide another good example. According to an Energy Information Administration (EIA) report, federal energy-specific subsidies and support to all forms of energy were estimated at $16.6 billion for fiscal year (FY) 2007. These EIA numbers for energy subsidies per megawatt hour (MWH) of energy break down as follows:

- Coal: $0.44/MWH
- Natural gas: $0.25/MWH
- Nuclear: $1.59/MWH
- Hydroelectric: $0.67/MWH
- Solar: $24.34/MWH
- Wind: $23.37/MWH

There is no doubt that changes in the distribution of subsidies by fuel type between 1999 and 2007 indicate a US governmental redirection of priorities and a greater support for renewable energy subsidies, a redirection paralleled in many parts of the world. In the United States, subsidies for renewables increased to 29 percent of total subsidies and support in 2007, up from 17 percent in 1999. In total dollars, this means that renewables accounted for $4.875 billion of the $16.581 billion in total 2007 energy subsidies. This means that $11.706 billion went to energy projects not related to renewable energy (EIA 2008, xii).

As such, it can be argued that the total dollars spent on traditional energy resources provide a greater market distortion than dollars spent on renewables. Thus the overall impact of dollars spent on traditional US fuel subsidies is greater than—more than twice—that spent on renewables. Providing subsidies for traditional fuel sources continues, even though companies that provide such sources are often well-established businesses that have (or should have) sound business models and methods. When such subsidies are used, the expectation is that more of whatever is being subsidized will result. Thus, subsidies for renewables should lead to more renewables, and subsidies for fossil fuels, however small, will also lead to more fossil fuels than would occur without the subsidies. This subsidization of fossil fuels serves to limit the effectiveness and slows the development of subsidies for sustainable energy sources, because the full benefit for sustainable sources is being offset by the support of more traditional energy sources.

Renewable energy sources account for a small percentage of total energy globally, as well as in the United States (7 percent in the United States for 2008), but that number is growing rapidly (EIA 2008, xii). If the stated goal is to produce more renewable energy, one option is to subsidize the cost. As a comparison, in the United States, the federal government subsidized nuclear power, and the investments (at least arguably) paid off. Those subsidy dollars spent in the 1950s are not reflected in today's subsidy costs, so nuclear power may appear significantly cheaper now than it is from a total cost perspective. Again, the nonsubsidy-related cleanup costs in places like Hanford, Washington—where nuclear waste has required billions of dollars of remediation efforts—have caused significant financial costs not usually added into today's costs.

The Future

Energy subsidies play, and will continue to play, a major role in global energy policies. There are significant tensions between the competing goals of current subsidies. Continuing subsidies for traditional fossil fuel sources serves to impede the progress of sustainable development, yet it may be essential for providing near-term access to energy for developing nations. Subsidies for renewable and sustainable sources provide incentives and opportunities for more environmentally friendly resources, yet misplaced incentives may improperly support resources that are not, and never will be, economically viable, thus impeding progress for other more promising sustainable energy sources.

There are some indications that the goals of sustainable development and economic development are becoming more possible. Countries, such as China, that have been reluctant to commit to sustainable practices and emissions reductions because of the potential negative impact on economic growth have nonetheless provided massive government subsidies to promote renewable energy manufacturing. Subsidies for renewable and sustainable energy will likely continue to increase as sustainable development becomes more obviously linked to economic development.

Energy subsidies for all energy sources are likely to be the norm for the foreseeable future.

Joshua P. FERSHEE
University of North Dakota School of Law

See also in the *Berkshire Encyclopedia of Sustainability* Climate Change Disclosure—Legal Framework; Climate Change Mitigation; Development, Sustainable—Overview of Laws and Commissions; Energy Conservation Incentives; Energy Security; Free Trade; Green Taxes; Investment Law, Energy; Utilities Regulation

Further Readings

Boaz, David. (Ed.). (2005). *Cato handbook on policy*. Washington, DC: Cato Institute.

Cooper, Christopher, & Sovacool, Benjamin K. (2007). Renewing America: The case for federal leadership on a national renewable portfolio standard (RPS). Retrieved September 2, 2010, from http://www.newenergychoices.org/dev/uploads/RPS%20Report_Cooper_Sovacool_FINAL_HILL.pdf

Energy Information Administration (EIA). (2008). Federal financial interventions and subsidies in energy markets 2007. Retrieved September 2, 2010, from http://www.eia.doe.gov/oiaf/servicerpt/subsidy2/index.html

Fershee, Joshua P. (2008). Changing resources, changing market: The impact of a national renewable portfolio standard on the US energy industry. *Energy Law Journal, 29*, 49–77.

Fershee, Joshua P. (2009). Atomic power, fossil fuels, and the environment: Lessons learned and the lasting impact of the Kennedy energy policies. *Texas Environmental Law Journal, 39*, 131–146.

Fershee, Joshua P. (2009). The geothermal bonus: Sustainable energy as a by-product of drilling for oil. *North Dakota Law Review, 85*(4), 893–905.

Kennedy, John F. (1964). Address at the University of North Dakota, 25 September 1963. In *Public papers of the presidents of the United States: John F. Kennedy, 1963* (pp. 715–719). Washington, DC: United States Government Printing Office.

Koplow, Douglas. (1996). Energy subsidies and the environment. In *Subsidies and environment: Exploring the linkages* (pp. 201–18). Paris: Organisation for Economic Co-operation and Development (OECD).

Kosmo, Mark. (1987). Money to burn? The high costs of energy subsidies. *World Resources Institute*. Retrieved October 6, 2010, from http://pdf.wri.org/moneytoburn_bw.pdf

Lieberman, Ben. (2010). Is wind the next ethanol? Retrieved November 5, 2010, from http://cei.org/sites/default/files/Ben%20Lieberman%20-%20Is%20Wind%20the%20Next%20Ethanol_0.pdf

Mann, Roberta F. (2009). Back to the future: Recommendations and predictions for greener tax policy. *Oregon Law Review, 88*(2), 355–404.

Mann, Roberta F., & Rowe, Meg. (forthcoming 2010). Taxation. In Michael B. Gerrard (Ed.), *The law of clean energy: Efficiency and renewables*. Chicago: ABA Publishing.

Ralls, Mary Ann. (2006). Congress got it right: There's no need to mandate renewable portfolio standards. *Energy Law Journal, 27*, 451–472.

Rural Electrification Act of 1936. (1936, May 20). Chap. 432, Title I, § 1, 49 Stat. 1363. (Current version at 7 U.S.C. § 901 [2006].)

Spence, David B. (2010). The political barriers to a national RPS. *Connecticut Law Review, 42*(5), 1451–1473.

Union of Concerned Scientists. (2009). Production tax credit for renewable energy. Retrieved September 19, 2010, from http://www.ucsusa.org/clean_energy/solutions/big_picture_solutions/production-tax-credit-for.html

Vandenbergh, Michael P.; Ackerly, Brooke A.; & Forster, Fred E. (2009). Micro-offsets and macro-transformation: An inconvenient view of climate change justice. *Harvard Environmental Law Review, 33*(2), 303–348.

von Moltke, Anja; McKee, Colin; & Morgan, Trevor (Eds.). (2004). *Energy subsidies: Lessons learned in assessing their impact and designing policy reforms*. Sheffield, UK: United Nations Environment Programme (UNEP) & Greenleaf Publishing.

Investment, CleanTech

CleanTech refers to technologies and business models that improve a product or a service's performance, productivity, or efficiency while reducing costs, using less energy and fewer materials, and lessening environmental damage. CleanTech investment in diverse sectors, including energy, water and wastewater treatment, manufacturing, advanced materials, transportation, and agriculture, provides competitive returns for investors and customers. Standardization, legislation, long-term planning, and incentives mediate its risks.

Since 2004, investment in the venture capital category known as CleanTech has emerged as a new trend that is fundamentally changing the face of the business of sustainability. Such technology—solar power, fuel cells, and aquaculture, for example—drives innovation as it has the potential to become an integral part of corporate value chains. (A "value chain" is a series of activities whose goal is to create value that exceeds the cost of providing a product or service, thereby generating a profit margin.) CleanTech refers to the "clean" technologies and business models that address the source of ecological problems caused by a product or a service's manufacturing or development process. It improves the product or service's performance, its productivity, or its efficiency as it reduces costs, uses less energy and fewer materials, and causes less environmental damage. Such technologies are clustered in diverse industry sectors, including energy (about two-thirds of total investment), water and wastewater treatment, manufacturing, advanced materials, transportation, and agriculture. The products or services cover a wide range including: energy-efficient lighting; wind and solar energy; water filtration; next-generation batteries; advanced materials that make products lighter, stronger, and/or cheaper; nontoxic pesticides; and enabling technologies that improve the performance of smart electricity grids.

Companies are realizing that integrating CleanTech solutions into their operations can potentially "green" their supply chains. It will also reduce their exposure to the uncertainties of climate change and water shortages and pollution, increase market valuations (the amount consumers are willing to pay), and, particularly, develop additional income streams in new markets. The challenge for corporations is to weigh the investment in CleanTech solutions, which often are capital intensive, against the solutions' potential long-term business value. For example, corporations and investors need to contend with the uncertainty of a technology's scalability or robustness and the need for new business models that would allow them to capture the value of the investment. They also need to consider the uncertainties of the financial markets that have hindered debt financing and public offerings, the volatility of oil prices, and the range of proposed policies aimed at adapting to climate change and reducing carbon emissions.

The CleanTech investment space is characterized by high capital investment, long technology-development cycles, a high degree of dependence on government policy, and (as of this early stage) uncertain exit strategies for investors. These characteristics contrast with typical information technology elements (e.g., materials in the solar value chain and smart grids that deliver electricity from alternative power sources), biotechnology (e.g., algae biofuels), and several other more mature investment domains. Hence new business models are needed for CleanTech investments, because investors and businesses that have transitioned from information technology or biotechnology into CleanTech have experienced a significant learning curve, particularly as it pertains to the valuation of companies, the size of investment rounds, and the potential for exit strategies (acquisitions or initial public offerings). The 2009 initial public offering of A123 Systems, which makes lithium-ion batteries that could be used in smart grids, potentially opened a window to access public markets profitably. Acquisitions, however, remain the most likely exit strategy.

Evidence is emerging that corporations are increasingly investing in CleanTech start-ups, joint ventures, and outright acquisitions, particularly in the energy, biofuel, and water-treatment technology areas, for the purpose of strategic differentiation and revenue growth. Examples include Exxon's $600 million investment in Synthetic Genomics during 2009 to develop superior algae strains for biodiesel production, General Electric's 2009 investment in A123 Systems, and Walmart's investment in energy- and water-efficiency companies to reduce carbon and water footprints (Cleantech Group 2009). (Carbon footprints and water footprints measure the amount of greenhouse gases created in the production of goods and services or the amount of freshwater used during daily activities, respectively.) The market for these technologies continues to expand, as more investors and corporations realize how sustainability-driven innovation through investment in CleanTech can better use natural resources in a way that provides economic value. What has fueled this trend, why are companies investing in CleanTech, and how is it related to leading innovations for corporate environmental sustainability?

Growth and Expansion

In the United States, CleanTech dates to the 1970s, when the environmental movement came of age after the Environmental Protection Agency was established. At that time, Congress enacted legislation such as the Clean Air Act and the Clean Water Act in response to public sentiment against environmental pollution and to the widespread evidence of environmental and public health hazards. These events and the oil crisis of 1973 prompted research and technology development for alternative energy generation, water-treatment processes, and "end-of-pipe" environmental treatment technologies that deal with pollution after it happens. Many of these technologies, however, were in an early stage of development, were too expensive, and did not have widespread political support; very few established companies embraced the innovative potential of this sector. This was due, in part, to the cost of these solutions and the absence of business models that would allow investors and companies to capture attractive returns. As a result, environmental technologies were often only implemented in small markets driven by regulatory compliance. Because of these events, the Cleantech Group (which coined and trademarked the term *cleantech*) argues that CleanTech should not be confused with the terms *environmental technology* or *green tech*, commonly used in the 1970s and 1980s.

To date, scores of companies and organizations in the United States and Europe are driving CleanTech business development and helping investors and corporations identify investment and acquisition opportunities. These include Clean Edge, a research and publishing firm dedicated to the CleanTech sector, which targets investors and entrepreneurs; Lux Research, an independent research and advisory firm, which provides corporations and investment funds with strategic advice and ongoing intelligence for emerging technologies; and Cleantech Europe, an advisory firm for entrepreneurs and investors, which canvasses the entire CleanTech space. Aside from business intelligence firms, the financial services industry has embraced CleanTech and focuses on the energy sector by establishing either new specializations within a firm or independent boutique operations.

Even at the start of the twenty-first century, the term *CleanTech* was not in the financial or business community's vocabulary. But since 2004, the sector has matured and gained recognition because it couples new CleanTech products or services with new business models that offer competitive returns for investors and customers. In recent years, this sector has seen a surge in financial innovations that drive businesses and consumers to adopt clean technologies. For example, rooftop solar technology is driven by long-term leasing and purchasing programs. In these programs, the technology is owned and maintained by the solar company, which also negotiates electricity rates with the utility to enable rebates to the consumer. These low-risk value propositions to consumers and peak power–demand mitigation values to utilities are becoming the mainstream for residential and commercial energy production and are being adopted in the waste-to-energy sector. A number of strategic drivers, which determine the success or direction of a company's business strategy, have spurred CleanTech's rapid growth. These drivers include the availability of private and public capital; the decreasing cost of technologies, which affects the scalability of solutions; the competition between governments to build jobs for the green economy; the certainty of climate change, which is influencing companies to disclose and mitigate their exposure risks; a changing consumer base that demands sustainable goods and services; and the resource demands of emerging economies such as India's and China's.

The best evidence that CleanTech has entered the mainstream is that governments around the world have made greening of the economy the centerpiece of their stimulus programs, at a cost that some estimate to be over $500 billion (Edenhofer and Stern 2009). This injection of government capital has accelerated investors' and corporations' interest, resulting in a green technology market rebound of

36 percent in the second quarter of 2009. At that time, more than three thousand venture-backed CleanTech companies were operating globally. Many more were funded through corporate investment, debt-equity financing, wealthy individuals, and government grants. Thus the intent of the technology, and the CleanTech venture based on it, is to mitigate carbon or water footprints through efficiency gains or other means because of their business value. Yet unintended consequences can result when forces other than markets pick which measures will be successes or failures. This has been the case with the ethanol biofuel mandates in the United States and the subsidies and preferential feed-in tariffs for solar and wind energy in Europe. Farmland gave way to energy farms, and food commodity prices spiked as a result. Hence the integration of policies and CleanTech solutions is awkward. Indeed, given the uncertainty of the outcome of the UN's 2009 Copenhagen climate discussions (COP-15), venture firms and entrepreneurs have been positioning themselves to become increasingly less reliant on environmental policies that may have unintended consequences or determine business value.

Impact on Sustainability

Corporations that make targeted investments in internal research and development, joint ventures, and acquisitions of companies that use disruptive technology (advances that improve a product or service in ways that are unexpected by the market) are creating both effective, innovative CleanTech solutions and value in their operations. This is where innovation in clean technologies and sustainability objectives intersect. The rationale is that technology investments help resolve the potential impacts of climate change and water risks on the long-term growth strategies (and thus market valuations) of corporations. Since 2004, disclosures in the financial market of economic, environmental, and social performance have become increasingly common in financial reporting and are important to organizational success. This is evident from corporations' disclosures in their Securities and Exchange Commission (SEC) filings of climate risks to their operations and supply chains and from their innovative solutions to reduce their exposure to the risk.

The Dow Jones Sustainability Indexes (DJSI) have tracked the financial performance of the leading sustainability-driven companies worldwide since 2000, and many organizations such as Ceres, the RiskMetrics Group, and the Carbon Disclosure Project analyze sustainability indicators. For example, Ceres is a national coalition of investors, environmental groups, and other public interest organizations that work with companies to address sustainability challenges such as global climate change. Ceres directs the Investor Network on Climate Risk, a group of more than seventy institutional investors from the United States and Europe that manage over $7 trillion in assets. A 2008 report published by Ceres describes how sixty-three of the largest consumer and technology companies across all industry sectors are positioning themselves to respond to the effects of climate change on their massive operations and supply chains (Risk Metrics 2008, 3). In the responses, the companies plan to reduce their carbon and water footprints. But because sustainability will increasingly become a driver for corporate strategy and differentiation, companies will need next-generation practices that change the existing business paradigms. To develop innovations that lead to next practices, executives must question the implicit assumptions behind their current practices. The Ceres report's recommended actions involve changing pay reward structures, governance systems, and supply chain management; setting renewable energy purchasing targets; and strategically investing in disruptive technologies. CleanTech allows businesses to change the way they operate, because technology differentiation results in strategic differentiation from the competition within and outside their industry sectors.

Many companies focus on the large portion of their carbon and water footprints that are in their supply chains. A number of leading companies began by managing their risks and developing standards to measure their emissions, and then moved forward to identify easy-to-achieve changes and the type of CleanTech investments to target. Consider Nike, whose extensive chain of footwear manufacturing sites accounts for 60 percent of its total carbon footprint. Because it is difficult to measure and control greenhouse gas emissions from raw materials processing, component suppliers, and the transportation of goods, these sites must collaborate with one another and with suppliers. Around 2009, Coca-Cola and Molson Coors started implementing a common industry standard to measure product lifecycle emissions, concentrating on managing their water footprints across the entire supply chain. Dell, Walmart, and several other companies directly engage suppliers in China to ensure that greenhouse gas emissions are assessed and reported.

But many companies go further. Walmart invested in CleanTech companies and technologies that offer solutions for greening their operations. For example, it implemented energy-efficient heating and lighting systems and pervious roofs and parking lot surfaces to restore the hydraulic cycle that moves water through land, the oceans, and the atmosphere. Walmart is even exploring colocating its warehouses with landfill gasification projects to form an off-grid power source. Energy companies such as Exxon and Chevron (alongside venture capital firms) invested heavily in algae biofuel start-up companies because they recognize that algae biodiesel technology may be scalable and produce an alternative to oil. Thus, they explored greening their product mix and tapped into new markets. These companies are particularly focused on the early stages in the value chain, such as the isolation, selection, and genetic engineering of highly efficient strains of algae, and the related extraction process technology. Engineering manufacturing companies such as Bosch and Siemens are diversifying in green technology by investing in start-ups across the solar and wind value chain and through acquisitions. Proponents of improving the corporate value chain believe that core engineering know-how can be directed to improve alternative energy and other technologies. This allows companies to tap into new markets while integrating the innovations into their operations and supply chains.

Risks and Controversies

The investment decisions in the growth and opportunities of CleanTech companies are at the center of a perfect storm: governments are unlocking unprecedented amounts of stimulus funds in order to green economies. Some argue that companies on the DJSI, compared with those not on the list, exhibit an increase of up to 15 percent in price-earnings ratios, indicating that the market values these companies more than it values companies that do not meet sustainability metrics. The market is looking at companies' climate risk disclosures in SEC filings. In addition, climate policies are influencing industry value chains, carbon markets for emissions trading, and consumer behaviors. Yet risks, unintended consequences, and other controversies may affect the future development of CleanTech innovations. Among these impacts are "greenwashing," or incorrectly stating the environmental benefits of a product, technology, or practice; the risk of governments awarding funds and determining policies based on a company's massive investment in targeted innovations; the risk of companies failing to make long-term plans; and the green paradox that policies aimed at curbing emissions may result in the acceleration of oil production.

How can companies and governments address these issues? First, some reporting standards for greening the supply chain and operations are emerging, and analysts are using them. As of the beginning of the twenty-first century, the implementation of climate and water risk reporting within companies is voluntary, and the metrics for various products and services are arbitrary. Increasing the standards of reporting and continuing to release disclosures will aid in the elimination of these risks. If the reporting of climate risks and the proposed mitigation strategies for companies, regardless of sectors, can be standardized, then corporations can hedge their financial risks to climate change and water risk uncertainties. For example, Swiss Re (a leading global reinsurer) and others are piloting weather insurance products that would reduce crop price volatility (affecting the food, beverage, and clothing industries) and transfer risk to financial players (the reinsurers or investors).

Second, the infusion of government capital will have the potential to remake alternative energy and other green technology businesses, influence investment returns, and affect industry value chains. For example, the United States's 2009 "green" economic stimulus plan targets a clean energy future by allotting $117.2 billion, or 12 percent of the $787.2 billion plan (Edenhofer and Stern 2009). It will achieve this through the development of plug-in hybrid cars and renewable energy technology, investment in energy efficiency, and a cap-and-trade program for trading pollution credits in order to reduce greenhouse gas emissions. China's government committed to a circular economy concept that would reduce, reuse, and recycle resources during manufacturing, transportation, and consumption by allotting $218 billion, or 33.4 percent of its budget. This legislation will allow Chinese auto manufacturers to leapfrog automotive technology by a generation and lead the green car revolution. It also identifies an official target for the capacity of installed solar and wind energy plants that well exceeds that of the United States. But history has shown that letting governments rather than markets decide what measures will be successful may be risky. For example, the US government credits for biofuel growers have pushed up commodity prices and caused a food-for-oil trade-off, affecting poor populations and increasing the water footprints of the biofuel industry. The emphasis on hybrid electric or all-electric cars, with their dependence on scarce supplies of lithium-ion batteries, leaves companies exposed to political risk and unsustainable extraction practices.

Third, companies must have a long-term plan that considers the opportunities and sustainability of their investments in CleanTech. The view that CleanTech is a mere add-on with a good cost-benefit ratio is shortsighted. These myopic views may result in companies shedding CleanTech when the market or policy incentives change, meaning the technology's social and environmental impacts will be small. The integration of CleanTech into the company's value structure has a greater potential to shift the company's competitive strategy and have a lasting impact for sustainability.

Finally, the so-called green paradox argues that as governments and companies strive to reduce emissions by reducing fossil fuel consumption (through CleanTech innovations in alternative energy, improved building insulation, and efficient cars), the global extraction of coal, gas, and oil will increase. The argument is that as companies green the economy, they exert a downward pressure on future fossil fuel prices (because less is needed). To maintain profits, owners of oil and gas fields will increase production, thus exacerbating climate change. The implication is that policies need to offer incentives for owners to leave supplies in the ground, rather than attempt to curb demand. The paradox is that curbs on demand have stimulated considerable CleanTech innovation. Taxation disincentives for owners are not politically viable, so a global carbon emissions trading system may be able to cap fuel consumption and slow down extraction rates. This would spur further innovations in carbon financing and bring financial and insurance services up the green value chain. (Carbon financing, as part of the Kyoto Protocol, generally refers to investments in greenhouse gas emission reduction projects and the creation of financial instruments that are tradable on the carbon market. The value chain essentially consists of three major players: project owners, traders or brokers, and buyers of carbon offsets. Even though not all greenhouse gas mitigation projects carry the same value, the carbon markets and their financial services players attempt to authenticate and verify the value of the offsets through financial instruments.)

Future Outlook

The fact that viable business models are generating attractive returns to investors and corporations shows that CleanTech is here to stay, regardless of the market, policy, and technology challenges it poses to corporate operations and value chains. While the United States's 2009 economic stimulus funding will be expended over two years, the impact in the markets of its green investment will take much longer to appear. The traction of SEC filings will continue as more companies participate in the disclosure of climate risks or participate in the Carbon Disclosure Project. With the standardization of measurements coming of age, companies will more easily measure their carbon and water footprints and identify opportunities to mitigate risks. Project financing and insurance pricing are increasingly tied to the climate exposures of corporations, and hence future favorable rates may induce companies to adopt carbon and water management strategies and consider integrating CleanTech solutions into their competitive strategies. In light of the maturation of the sector since 2004, which was driven by venture capital and private equity, the early twenty-first century will be driven by growth and expansion in the mainstream economy that is fueled by government programs. It is, however, important to differentiate the outlook by technology domain.

Energy will continue to be the main sector for investment, particularly transportation (batteries and fuel cells), biofuels (algae and noncrop plants), and the continued expansion of solar and wind power. In the third quarter of 2009, CleanTech investments for the first time exceeded those in software, biomedical devices, and biotechnology, with 25 percent of the total venture investment occurring in the United States (Cleantech Group 2009, 13). Solar is the top sector, with a 28 percent share of investment, closely followed by transportation (25 percent) and biofuels (9 percent) (Cleantech Group 2009, 10). In the short term, analysts expect major consolidation to occur in the value chains for these industries. The causes are the role of government, private capital shortages, overcapacity in the face of lower energy demand, and decreasing prices for the integrated technology. This consolidation will stimulate investment in energy companies because investors can control the whole supply chain and respond to US mandates under the stimulus package. This trend is already evident for the solar industry and is likely to follow in the wind industry. New financial models and incentives will be needed to ensure the stability of this CleanTech industry. For example, utilities pay feed-in tariffs, or set prices for paying the end users who provide solar power to the power grid. Making these tariffs consistent across all solar projects (as is proposed in China) will reduce costs and make the projects viable, thereby allowing companies to calculate risks and returns. Even though it is hard to project where investments are headed, findings from Deloitte Touche Tohmatsu's 2009 *Global Trends in Venture Capital* report shows that 63 percent of venture capitalists around the globe intend to increase their exposure to the CleanTech category over the next three years—a far higher percentage than any other sector. The trend looks set to continue for some time.

The demand for renewable energy by utilities is growing, as they must comply with state (and most likely federal) renewable portfolio standards (RPSs) by 2020. The doubts some have that the widespread, integral use of solar power in the RPSs are waning. (Even in the oil- and gas-intensive city of Houston, Texas, utilities are buying solar technology.) Global competition is playing a role as well, with capacity targets for installed alternative energy systems in China rapidly eclipsing those in the United States. A major challenge, aside from policy incentives, is the creation of a smart grid that can handle the highly variable production of electricity from renewable energy sources, one that will allow for storage and can track the "green" and "brown" electrons. A smart grid links power production, transmission, and distribution from centralized (e.g., coal-fired power plants) and distributed sources (e.g., wind farms or electric vehicles) with consumer demand. With electricity being generated from traditional coal or oil plants (brown electrons) as well as from the wind, solar energy, biofuels, and batteries (green electrons), an intricate management system needs to be developed. In the United States, many start-up companies in all parts of the value chain are working together to develop enabling technologies and management aspects that meet a federal mandate to create a smart grid. Considering the fixed infrastructure of the current grid, companies will likely locate green smart grids along corridors (e.g., in Arizona and California) to test their feasibility. Bond measures will likely fund their construction in a piecemeal fashion, and business models will be developed to monetize them with attractive returns.

In other sectors, advances in battery technology aside from lithium-ion batteries will fuel the generation of cars that will be developed in the second decade of the century. But first, battery costs must come down and their efficiency must improve. Experience with hybrid technology and General Motors's Volt electric car will prove invaluable to assess market demand for this technology and to drive innovations in electrical grid infrastructure. Experience with hybrid and electric technology might also affect business models pertaining to the purchase, maintenance, and afterlife care of the battery system. Corporations may be highly motivated to move in this direction because of their investments in fuel-efficient fleets, whether on an ownership, lease, or joint venture basis. Future fuel prices, the cost of technology, and government incentives will play a major role in the adoption of this technology. Considering the long development cycles and fragmentation of the industry, these innovations will presumably have a long time on the horizon before they play out in the marketplace.

Finally, conservative deregulation in the water industry sector has seen much innovation and investment or acquisition activity, both for ultrahigh-purity and high-volume applications. For example, General Electric acquired Canada's Glegg Water Company, which developed a superior electrodeionization technology and product (the E-Cell) for ultraclean water treatment. Targeting high-value customers such as the pharmaceutical and semiconductor industries, General Electric's Water and Process Technologies division recognized the opportunity as an acquisition and business opportunity. Two main strategic drivers will increase investments in water technologies: SEC disclosures of corporate water footprints and the energy–water nexus. This nexus includes CleanTech innovations that address the energy used for water conveyance and treatment (e.g., to operate filtration systems, microturbines, and fuel cells). Other innovations address water use for energy production from coal, gas, nuclear, biofuel, and utility-scale solar sources. This water industry sector is moving up the value chain due to a number of factors. First, climate change affects corporate risks to water availability and quality. Second, water costs for corporate users are being renegotiated, and finally, multistakeholder water use is starting to drive new legislation.

Peter ADRIAENS

Ross School of Business, University of Michigan

See also in the *Berkshire Encyclopedia of Sustainability* Biomimicry; Cap-and-Trade Legislation; Climate Change Disclosure; Energy Efficiency; Energy Industries—Overview of Renewables; Green-Collar Jobs; Investment, Socially Responsible (SRI); Product-Service Systems (PSSs); Risk Management; Supply Chain Management; True Cost Economics

Further Reading

Alter, Alexandra. (2009, February 17). Yet another footprint to worry about: Water. Retrieved November 16, 2009, from http://online.wsj.com/article/SB123483638138996305.html

Carbon Disclosure Project. (2009). Retrieved November 16, 2009, from http://www.cdproject.net/

Clean Edge. (2009). Retrieved November16, 2009, from http://www.cleanedge.com/

Cleantech Group. (2009). *Cleantech investment monitor: Third quarter 2009.* Retrieved November 16, 2009, from http://www.deloitte.com/view/en_US/us/Services/additional-services/Corporate-Responsibility-Sustainability/clean-tech/article/46dd16e228725210VgnVCM100000ba42f00aRCRD.htm

Deloitte Touche Tohmatsu. (2009, June). *Global trends in venture capital 2009 global report.* Retrieved December 8, 2009, from http://www.deloitte.com/view/en_GX/global/article/e79af6b085912210VgnVCM100000ba42f00aRCRD.htm

Edenhofer, Ottmar, & Stern, Nicholas. (2009). *Towards a global green recovery: Recommendations for immediate G20 action.* Retrieved December 14, 2009, from http://www2.lse.ac.uk/granthamInstitute/publications/GlobalGreenRecovery_April09.pdf

Esty, Daniel C., & Winston, Andrew S. (2006). *Green to gold: How smart companies use environmental strategy to innovate, create value, and build competitive advantage.* New Haven, CT: Yale University Press.

Kammen, Daniel M. (2006). The rise of renewable energy. *Scientific American, 295*(3), 85–93.

Metcalfe, Robert M. (2008, September). Learning from the Internet. Retrieved November 16, 2009, from http://www.sciamdigital.com/index.cfm?fa=Products.ViewIssue&ISSUEID_CHAR=7642E940-3048-8A5E-103FDB5762BAF1D5

Pernick, Ron, & Wilder, Clint. (2008). *The clean tech revolution.* New York: Harper Collins Business.

The power and the glory. (2008, June 19). Retrieved November 16, 2009, from http://www.economist.com/specialreports/displaystory.cfm?story_id=11565685

RiskMetrics Group, & Ceres. (2008, December). *Corporate governance and climate change: Consumer and technology companies.* Retrieved November 16, 2009, from http://www.ceres.org/Document.Doc?id=398

Investment Law, Energy

Countries worldwide have sovereignty over their natural resources and may develop them according to their own environmental laws and regulations. International law recognizes the need for modern investment treaties—such as the North American Free Trade Agreement and the Energy Charter Treaty—that include provisions affecting the energy industry and its sustainability. Assessing how such treaties will impact the foreign investment climate in participating states remains a challenge for the future.

It is widely accepted that foreign investment is critical for sustainable development, and more so with respect to the oil and gas industry, a capital-intensive industry with a long gestation period (Elder 1991; Ginther, Denters, and de Waart 1995; Vale Columbia Centre and WAIPA 2010). Although international law recognizes national sovereignty over natural resources, international law places limitations on the exercise of such sovereign rights in a situation where it causes harm to other states or when it is pursued in an unsustainable manner (Cameron 2010). Consequently, the emergence of sustainability laws has had an impact on worldwide energy investments (Sands 1995).

The development and use of oil, gas, and nuclear energy are subject to specific international regulation. Other sources of energy—coal and renewable sources such as wind, solar, and geothermal—are scarcely regulated by international instruments. Because the effects of their operations are thought to be confined within national borders, they constitute an insignificant percentage of the global energy mix, and/or they cause relatively less damage to the environment as to attract much international attention. The development and use of oil, gas, and nuclear energy pose more serious environmental, health, social, and cultural consequences; they may result in oil spills, pollution and degradation of the environment, displacement of local communities, or accidents (Smith and RMMLF 2010; Park 2002; Gao 1998; Horbach 1999).

The North American Free Trade Agreement (NAFTA) and the Energy Charter Treaty (ECT), two multilateral trade and investment treaties, have direct relevance to the energy sector and how it might be developed in a sustainable manner. Other relevant international instruments include the Law of the Sea Convention 1982, Climate Change Convention 1992, the Convention on Biodiversity 1992, the United Nations Framework Convention on Climate Change (UNFCC) 1997, the Kyoto Protocol 1997, and the increasing network of bilateral investment treaties (Cameron 2010; Vandevelde 2010). This article focuses only on NAFTA and the Energy Charter Treaty because, generally speaking, the key provisions on investment in both treaties are similar to those contained in most other investment treaties, and, more specifically, both contain explicit provisions on sustainability. Additionally, the ECT deals exclusively with the energy sector, and NAFTA contains a chapter on energy.

Historical Background and Objectives

After fourteen months of negotiations, the leaders of the United States, Canada, and Mexico signed NAFTA on 7 October 1992, and the treaty came into effect on 1 January 1994. NAFTA was built upon the Canada-United States Free Trade Agreement of 1989. Before the NAFTA agreement came into effect, however, a Supplementary Agreement was reached in August 1993, upon the insistence of President Clinton, to address concerns over protection of the environment and workers' rights, neither of which were sufficiently addressed in the original NAFTA. The

Supplemental Agreement is a comprehensive trade and investment agreement that affects all aspects of doing business in Canada, Mexico, and the United States. It seeks to eliminate barriers (such as tariffs) to the free flow of goods and services, removes restrictions to investment, and strengthens intellectual property rights among the three countries. It was envisaged by the contracting states as an effort toward establishing a free trade area in the North American continent (similar to the European Union and the European Free Trade area) so as to enhance the competitiveness of the region in global trade and investment (Smith and Cluchey 1994). For the Canadian government, the main objectives for signing NAFTA were to get Canadian goods, services, and capital access to Mexico on an equal footing with the United States, and to make Canada attractive to foreign investors wishing to invest in the North American market (Saunders 1994). From the United States' perspective, NAFTA would help provide closer and more stable sources of energy supplies from Canada and Mexico and reduce its over reliance on the more unstable Middle Eastern oil (Smith and Cluchey 1994).

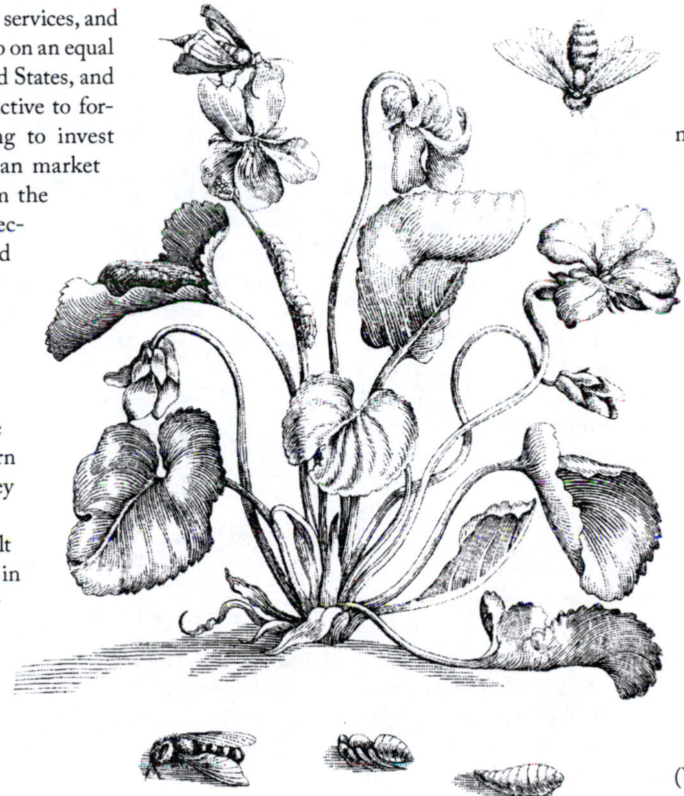

The ECT is the result of a political initiative in Europe in the early 1990s following the end of the Cold War and the disintegration of the Soviet Union. The then Dutch prime minister initiated the process by suggesting the creation of a European energy community and urged the development of the European Energy Charter, a nonbinding political declaration signed in The Hague in 1991 by fifty-six states, including the European Community and Australia, where ratification is still in progress. The Charter "represents a political commitment to cooperate in the energy sector, based on the principles of development of open and efficient energy markets" (Corell 2005). The participants to the Charter did acknowledge the need for a binding international legal framework for effective cooperation in the sector, and so negotiations for the ECT started in late 1991. The ECT and the Protocol on Energy Efficiency and Related Environmental Aspects (PEEREA) were signed in Lisbon on 17 December 1994 and came into force on 16 April 1998. Although the United States and Canada participated in the negotiations, they did not sign the treaty. As the name suggests, the ECT is a sectoral agreement; it deals solely with the energy industry and covers issues of trade, investment, transit, and environment.

The main goals of the initial negotiation parties were to help the Eastern European countries transition to market economies by injecting Western investment into their energy sector, which would help in ensuring security of energy supplies to western Europe. In this regard, the ECT "plays an important role as part of an international effort to build a legal foundation for energy security, based on the principles of open, competitive markets and sustainable development" (Energy Charter Secretariat n.d.). For the Eastern European and resource countries, "the main attraction of the treaty was to appear attractive, to be seen to play the rules of the global economy, reduce their political risk perception and not to be left out of possibly significant energy policy dialogue" (Wälde 2004). Overall, the "fundamental aim of the ECT is to strengthen the rule of law on energy issues by creating a level playing field of rules to be observed by all participating governments, thereby mitigating risks associated with energy-related investment and trade" (Energy Charter Secretariat n.d.). To date, the ECT has been signed by fifty-four members while twenty-four countries act as observers.

Energy Investment

With regard to energy, NAFTA Article 602 states that the agreement "applies to measures relating to energy and basic petrochemical goods originating in the territories of the Parties and to measures relating to investment and to the cross-border trade in services associated with such goods." With respect to the Mexican energy sector, NAFTA falls short of achieving the objective of creating a common energy market and the liberalization of the sector. Due to historical and constitutional reasons, limitations are placed on foreign involvement in Mexico's energy sector, and these policies are reflected in NAFTA. Chapter 6, Annex 602.3 reserves to the Mexican state or its state entities all activities relating to exploration and exploitation of oil and gas, ownership and operation of pipelines, all foreign trade, transportation, storage, and distribution of Mexican crude oil and natural gas. Thus, NAFTA may have less impact on energy trade and investment in Mexico than it does in the United States and Canada, which "are bound to permit the free flow of energy goods and investments and of services throughout the energy sector" in accordance with the agreement (Herman 1997). This has the effect of constraining the ability of future Canadian governments from reverting to the old nationalistic and protectionist energy policy of the 1980s as reflected in the National Energy Programme of 1980 (Saunders 1994). Even with regard to Mexico, there are possibilities for foreign investment in the nonbasic petrochemicals and certain aspects of electricity generation sectors, which are not subject to the constitutional restriction on investment (Saunders 1994). Energy investment under NAFTA chapter 6 is reinforced by chapter 11, which is the most important chapter because it defines the rights and obligations of investors and the state parties. Similar to other investment treaties, including the ECT (Part III, Articles 10–17), NAFTA chapter 11 provides for a definition of protected "investment" and "investors," and the standard of treatment to be accorded such investments and investors from other member states. The standards of treatment include national and most-favored-nation treatment, fair and equitable treatment, full protection, and security. Other provisions that address more specific situations include conditions on expropriation of covered investment, guarantees on rights of free transfer of payments related to an investment, prohibition on performance requirements, and provisions intended to promote transparency or access to courts. These substantive provisions are backed by investor-state or state-state dispute resolution provisions. The first provision vests in the foreign investor a direct action (through international arbitration) against the host state for alleged violation of the investor's substantive rights, and the second includes a dispute settlement mechanism between the two state parties concerning the interpretation or application of the treaty (Vandevelde 2010; Konoplyanik and Wälde 2006). Over the years, several cases have been brought by foreign investors seeking to challenge measures adopted by host states that were alleged to not conform to their international investment treaty obligations under the applicable treaty (Vandevelde 2010; Salacuse 2010).

Sustainable Development

Both NAFTA and the ECT contain provisions on sustainability. In its preamble, the NAFTA state parties signal their support to promote sustainable development. In order to achieve this objective, the treaty permits each contracting state to take appropriate measures to ensure that investment activity in its territories is implemented in a manner consistent with environmental protection, provided such measures are consistent with the overall objectives of NAFTA. Furthermore, as specified by Article 1114(2), "it is inappropriate to encourage investment by relaxing domestic health, safety, or environmental measures."

In addition, the NAFTA Side Agreement on Environmental Cooperation provides for sanctions for lax enforcement of domestic environmental laws and standards. The provisions are to ensure that trade and investment activities in the member states' territories are conducted in a sustainable manner by preventing a "race to the bottom" approach by member states. Thus the Side Agreement (Article 1) expresses the willingness of the parties to promote certain conditions:

- sustainable development based on cooperation and mutually supportive environmental and economic policies
- enhanced compliance with and enforcement of environmental requirements
- transparency and public participation in developing environmental norms
- economically effective environmental measures

Under the Side Agreement, members of the public, including nongovernmental organizations (NGOs), are allowed to challenge a state party if it fails to effectively enforce its environmental laws or regulations. The Side Agreement establishes the Commission for Environmental Cooperation (CEC), the institutional framework to receive petitions from members of the public and, if necessary, investigate the claims and prepare a factual record, which might be published by the Council. (The Council is the CEC's governing body, composed of high-level environmental authorities from Canada, Mexico, and the United States.) The process has been utilized by many individuals and organizations with varying degrees of success. Hence, it has been described as a "'spotlighting' instrument intended to enhance governmental accountability and transparency" (Markell 2010; Knox 2010).

Similarly, Article 19 of the ECT enjoins member states to "strive to take precautionary measures to prevent or minimize environmental degradation" and to "take account of environmental considerations throughout the formulation and implementation of their energy policies." The treaty also requires that the member states take specific actions relating to the following:

- the promotion of market-based price reform and fuller reflection of environmental costs and benefits
- the encouragement of international cooperation
- information sharing on environmentally sound and economically efficient energy policies
- the promotion of environmental impact assessment activities and monitoring
- the promotion of public awareness on relevant environmental programs
- the research and development of energy efficient and environmentally sound technologies, including the transfer of technology

Although these are soft law and not legally binding obligations, they may have indirect legal implications, such as "justifying regulatory measures subject to the scrutiny of the investment protection regime" (Wälde 2001). Furthermore, the Protocol on Energy Efficiency and Related Environmental Aspects (PEEREA) requires member states to formulate policy principles aimed at improving energy efficiency, reducing negative environmental impact, and fostering international cooperation between member states. The implementation of the PEEREA would provide transition economies with good practices and an opportunity to share experiences and policy advice on energy efficiency issues with their Western counterparts.

Impact and Current Challenges

According to the United States government, NAFTA has achieved its core goals of expanding trade and investment between the three countries. It asserts that "from 1993 to 2007, trade among the NAFTA nations more than tripled, from $297 billion to $930 billion" and that "business investment in the United States has risen by 117 percent since 1993, compared to 45 percent increase between 1979 and 1993" (Office of the United States Trade Representative 2008). With respect to the energy sector, the business and legal climate in Canada and Mexico have become more hospitable to foreign energy-related investment as a result of NAFTA (Smith and Cluchey 1994). Concerning sustainable development, it has been noted that the NAFTA regime "has had its greatest success as a regional effort to promote sustainable development. It has contributed to stronger environmental protections, especially in Mexico," and it has formed a basis for subsequent United States Free Trade Agreements to include environmental protection provisions (Knox 2010). But concerns over the fairness or neutrality of the process, the slow pace of the procedure, and the apparently "toothless" character of the mechanisms have marred the success of the Side Agreement on Environment (Markell 2010; Knox 2010). Furthermore, one of the strongest challenges facing NAFTA, the ECT, and other investment treaties relates to how to reconcile the obligations of the state parties toward foreign investors and the needs for regulatory autonomy in the areas of environmental protection and human rights. The absence of clear guidelines in the investment treaties on how to resolve such potential conflicts poses a serious legal and policy challenge to state parties and foreign investors who have to rely on the interpretative decisions of arbitral tribunals (Kingsbury and Schill 2010; Vandevelde 2009).

Some general conclusions can be drawn from this overview of energy investment law. First, although every country has sovereignty over its natural resources and the right to develop them in accordance with its environmental laws and regulations, modern investment treaties such as NAFTA and the ECT may constrain a member-state's discretion. Second, modern investment treaties vest substantive and procedural rights in foreign investors to challenge egregious host-state measures before international tribunals in a manner never before known under general international law. Third, it is not yet settled how to strike a proper balance between energy investment and sustainable development, hence the uncertainty in the legal relationship between foreign investors and host states. Finally, it is difficult to assess the extent to which modern investment treaties, such as NAFTA and the ECT, have contributed to improving the investment climate in the energy sector of the state parties.

Peter CAMERON and Abba KOLO
University of Dundee

See also in the *Berkshire Encyclopedia of Sustainability* Climate Change Disclosure—Legal Framework; Climate Change Mitigation; Development, Sustainable—Overview of Laws and Commissions; Energy Conservation Initiatives; Environmental Law—United States and Canada; Environmental Law, Soft vs. Hard; Free Trade; Investment Law, Foreign; Green Taxes; Utilities Regulation

Further Readings

Cameron, Peter. (2010). *International energy investment law: The pursuit of stability*. Oxford, UK: Oxford University Press.

Corell, Hans. (2006). Introduction to the Energy Charter Treaty. In Clarisse Ribeiro (Ed.), *Investment arbitration and the Energy Charter Treaty* (pp. 1-3). Huntington, NY: JurisNet.

Elder, P. S. (1991). Sustainability. *McGill Law Journal, 36*, 831–834.

Energy Charter Secretariat. (n.d.) About the charter. Retrieved November 23, 2010, from http://www.encharter.org/index.php?id=7

Gao, Zhiguo. (1998). Environmental regulation of oil and gas in the twentieth century and beyond: An introduction and overview. In Zhiguo Gao (Ed.), *Environmental regulation of oil and gas* (pp. 3–58). London: Kluwer Law International.

Ginther, Konrad; Denters, Erik; & de Waart, P. J. I. M. (1995). *Sustainable development and good governance*. Dordrecht, The Netherlands: M. Nijhoff.

Herman, Lawrence L. (1997). NAFTA and the ECT: Divergent approaches with a core of harmony. *Journal of Energy and Natural Resources Law, 15*, 129–133.

Horbach, Nathalie. (1999). Lacunae of international nuclear liability agreements. In Nathalie Horbach (Ed.), *Contemporary developments in nuclear energy law* (p. 43). London: Kluwer Law International.

Kingsbury, Benedict, & Schill, Stephan. (2010). Public law concepts to balance investors' rights with state regulatory actions in the public interest – the concept of proportionality. In Stephan Schill (Ed.), *International investment law and comparative public law* (pp. 75–104). Oxford, UK: Oxford University Press.

Knox, John H. (2010). The neglected lessons of the NAFTA environmental regime. *Wake Forest Law Review, 45*, 391–424.

Konoplyanik, Andrei, & Wälde, Thomas. (2006). Energy Charter Treaty and its role in international energy. *Journal of Energy and Natural Resources Law, 24*(4), 523–558.

Markell, David. (2010). The role of spotlighting procedures in promoting citizen participation, transparency, and accountability. *Wake Forest Law Review, 45*, 425–467.

Office of the United States Trade Representative. (2008). NAFTA facts. Retrieved November 23, 2010, from http://www.ustr.gov

Park, Patricia. (2002). *Energy law and the environment*. London: Taylor & Francis.

Salacuse, Jeswald W. (2010). *The law of investment treaties*. Oxford, UK: Oxford University Press.

Sands, Philippe. (1995). International law in the field of sustainable development: Emerging legal principles. In Winfried Lang (Ed.), *Sustainable development and international law* (pp. 53–66). London: Graham & Trotman / M. Nijhoff.

Saunders, J. Owen. (1994). GATT, NAFTA and the North American energy trade: A Canadian perspective. *Journal of Energy and Natural Resources Law, 12*, 4–9.

Smith, Ernest E. & Cluchey, David P. (1994). GATT, NAFTA and the trade in energy: A US perspective. *Journal of Energy and Natural Resources Law, 12*, 27–58.

Smith, Ernest E., & Rocky Mountain Mineral Law Foundation (RMMLF). (2010). *International petroleum transactions* (3rd ed.). Denver, CO: Rocky Mountain Mineral Law Foundation.

Vale Columbia Centre & World Association of Investment Promotion Agencies (WAIPA). (2010). Investment promotion agencies and sustainable FDI: Moving towards the fourth generation of investment promotion. Retrieved November 23, 2010, from http://www.vcc.columbia.edu/files/vale/content/IPASurvey.pdf

Vandevelde, Kenneth. (2009). A comparison of the 2004 and 1994 US model BITs. In Karl Sauvant (Ed.), *Yearbook on international investment law & policy 2008–2009* (pp. 283–315). Oxford, UK: Oxford University Press.

Vandevelde, Kenneth. (2010). *Bilateral investment treaties*. Oxford, UK: Oxford University Press.

Wälde, Thomas. (2001). International disciplines on national environmental regulation: With particular focus on multilateral investment treaties. In Permanent Court of Arbitration (Ed.), *International investments and protection of the environment* (pp. 29-47). The Hague, The Netherlands: Kluwer Law International.

Wälde, Thomas. (2004). The Energy Charter Treaty: Expanding the liberalization of energy industries. Unpublished paper on file with the authors.

Iron Ore

Iron ore is the key ingredient for steel, a metal critical to modern infrastructure and technologies. Iron is abundant, but deposits of high concentration and large size are needed to allow economical mining. Sustainability of iron production will require continuing development in such areas as transport logistics, energy options for mining and transport, and responsible management of land and water resources.

Iron ore is the primary mineral used to produce iron and steel—both of which are crucial to infrastructure, technology, materials consumption, economic development, and social progress. Steel consumption per person is a common indicator used to assess the extent of economic progress, and, indirectly, thereby the development status of a country or region. Iron ore and steel and its associated environmental issues are therefore critical in understanding and assessing the sustainability of natural resources.

Iron Ore Resources and Production

Iron (Fe) is relatively abundant in the Earth's crust in a wide variety of minerals, but only a few are of economic interest in the production of iron and steel goods. The three dominant minerals are hematite (Fe_2O_3), goethite ($FeO(OH)$), and magnetite (Fe_3O_4), with minor production also sourced from pyrite (FeS_2) in some regions. Typical concentrations of economic ore are 40–60 percent iron.

The estimates of minable iron ore resources vary markedly, depending on the economic assumptions used (see USGS 2011 for some recent estimates). Key factors include mineralogy (especially hematite versus magnetite), ore grade (percent iron), deleterious impurities, total deposit size, infrastructure, and environmental issues. The countries with the largest resources are Australia, Brazil, and Russia, with significant resources known in Ukraine, China, South Africa, and the United States. The iron content of world reserves is estimated to be at least 87 billion tonnes, with further amounts known but of uncertain economic character. Given the surging demand from Asia, especially China, exploration has increased dramatically in recent years, leading many countries around the world to increase their economic reserves (especially Australia, India, and Brazil).

Typical iron ore mines are based on large deposits mined by open-cut methods. The four biggest iron ore producers are China, Australia, Brazil, and India, collectively representing over 80 percent of annual iron production worldwide. Although China produces twice as much as Australia and Brazil, it is from much lower-grade ores, and China is increasingly looking to imports to maintain steel production and meet demand. In terms of world exports, Australia and Brazil dominate the seaborne trade in iron ore, especially to the large steel mills of Japan, China, and South Korea. In 2010, iron ore production was about 2.4 billion tonnes worldwide.

Given iron ore's relatively straightforward mining and processing, technical challenges revolve around transport logistics, especially long-distance railways and export ports on coasts. This generally favors large, long-life iron ore operations—such as the Pilbara region of Western Australia or Carajas in Brazil—which can maintain lower unit costs of production. Another major issue is that magnetite ores are commonly lower grade and require significant processing to form iron ore pellets, leading to a preference for direct shipping or simple hematite ores.

Sustainability Issues

In the twenty-first century, a myriad of complex issues will affect the sustainability of iron ore resources. First, the mining and transport of iron ore presently require fossil fuels to run the shovels, trucks, and trains. Given the growing concerns about reaching what is called *peak oil* (the limit of global oil production), there is a pressing need to look at long-term energy options, such as biodiesel fuel or electrification. Second, with deeper mines, longer haul distances, more processing, and the need to address impurities, there is significant upward pressure on the carbon intensity of iron ore production. The generation of greenhouse gas emissions by iron ore production will be increasingly scrutinized, in accord with global desires to reduce such emissions and avert the impacts of climate change.

Managing impacts on water resources and the land surrounding mines remains an ongoing challenge. Iron ore mines operate on a large scale and can affect regional landscapes and associated surface water and groundwater resources. In Australia, Brazil, and India, as well as emerging projects in Africa, almost all iron ore mines are in tropical climates, where water management is a critical focus because of flooding and pollution risks.

Given typical grades of 40–60 percent iron in raw ore and processed concentrate grades of about 60–65 percent iron, most ore becomes concentrate and is smelted to produce steel. As such, only a small fraction becomes tailings (less than 20 percent), although lower grade magnetite projects will generate a tonne of tailings for every tonne of iron ore pellets. In any case, the large tonnage of iron ore mined and processed annually leads to substantial volumes of tailings being produced globally. Tailings dams are the dominant form of management, although in Canada some iron ore mines deposit tailings in lakes.

An area of growing knowledge in iron ore mining concerns the risks from acid mine drainage (AMD). Although the dominant economic iron minerals are oxides or hydroxides, the rock surrounding the iron ore can contain pyrites (iron sulfide), which will react with water and oxygen when exposed at the surface to form sulfuric acid, in turn dissolving other salts and heavy metals to form AMD. For example, in the Pilbara iron ore district of Western Australia, significant AMD risks are associated with the fine-grained shales, waste rock that requires active planning, monitoring and management (see Waters and O'Kane 2003). The presence and risks of AMD will be site specific but it cannot be ignored given the large scale of impacts should it occur.

Economic flows from the magnitude of the iron ore industry are currently the subject of considerable debate, especially in existing producer countries such as Australia, but also in potential new producing countries in Africa. Controversial subjects, critical to the economic sustainability of nations, include the sharing of economic benefits, taxation rates, and the possibility of banking part of the economic flow into a sovereign wealth fund to build financial capital.

The long-term trajectory for developed nations has been toward recycling of steel, with most steel now produced by processing recycled steel in an electric arc furnace (EAF) (see World Steel Association 2007). The use of EAF technology makes steel production considerably lower in carbon intensity, and recycling steel takes pressure off the production of additional iron—both outcomes beneficial to sustainability.

The future of the global iron ore industry can be assured for some decades, given that existing ore reserves and mineral resources are considerable and continue to grow. A range of immediate and medium-term issues, however, will affect the iron ore sector, such as peak oil and energy security, greenhouse gas emissions and climate change policies, economic and taxation policies, as well as water resources management challenges.

Gavin M. MUDD
Monash University

See also in the *Berkshire Encyclopedia of Sustainability* Aluminum; Chromium; Coltan; Copper; Electronics—Raw Materials; Gold; Heavy Metals; Iron Ore; Lead; Lithium; Minerals Scarcity; Mining—Metals; Nickel; Platinum Group Elements; Rare Earth Elements; Recycling; Silver; Thorium; Tin; Titanium; Uranium

Further Reading

International Institute for Environment and Development (IIED), & World Business Council for Sustainable Development (WBCSD). (2002). *Breaking new ground: Mining, minerals and sustainable development.* London: Earthscan Publications for IIED and WBSCSD.

Raymond, Robert. (1984). *Out of the fiery furnace: The impact of metals on the history of mankind.* Melbourne, Australia: Macmillan.

United States Geological Survey (USGS). (2011). Minerals commodity summaries 2011. Reston, VA: author. Retrieved September 29, 2011, from http://minerals.usgs.gov/minerals/pubs/mcs/

Waters, P., & O'Kane, Michael. (2003). Mining and storage of reactive shale at BHP Billiton's Mt. Whaleback Mine. Proceedings of the Sixth International Conference on Acid Rock Drainage (6th ICARD), T. Farrell & G. Taylor (Eds.), Australasian Institute of Mining & Metallurgy (pp 155–161). Cairns, Queensland, Australia.

World Steel Association. (2007). Steel statistical yearbook 2007. International Iron & Steel Institute (IISI) Committee on Economic Studies, World Steel Association (WSA). Brussels: author.

Yellishetty, Mohan; Mudd, Gavin M.; & Ranjith, P. G. (2011). The steel industry, abiotic resource depletion and life cycle assessment: A real or perceived issue? *Journal of Cleaner Production, 19*(1), 78–91).

Lighting, Indoor

Indoor lighting is a major agent of energy consumption and of carbon dioxide emissions. Technological substitution of incandescent lights with energy efficient alternatives has helped to improve lighting efficiency, but global lighting demand is still on an unsustainable upward trajectory. Along with improving technological efficiency, cultural lighting preferences and habits must be addressed in the effort to transform human dependency on energy-intensive lighting practices.

Lighting is vital to human sustenance. It enables us to perform daily tasks and enhances our mood. The quality of an indoor space and the well-being of those who occupy it can vary greatly depending on lighting conditions. Natural daylight is an important means of maintaining our biological rhythm and of marking important daily events (e.g., dawn, noon, or night). Artificial light can transform a dark and oppressive place into a stimulating one and is vital to contemporary society, where the working day extends long past hours of natural daylight. But artificial lighting is also associated with physical and visual discomfort, and prolonged exposure can be detrimental to human health.

Indoor lighting is also a major source of energy consumption and of carbon dioxide emissions. Electric lighting consumes 19 percent of total global electricity production (IEA 2006, 25). In 2008, lighting accounted for 15.4 percent of residential electricity consumption in the United States (US EIA 2008). The International Energy Agency (IEA) has predicted that without rapid action the amount of energy used for lighting globally will be 80 percent higher in 2030 than in 2006 (IEA 2006, 26). Significant growth in energy demand for lighting has already been witnessed in developed countries like the United Kingdom, where it increased by 63 percent between 1970 and 2000 (DTI 2002, 26). In rapidly developing countries like China and India, the combination of expectations for improved lighting quality, along with growth in population and building construction, is anticipated to significantly increase the rate of global lighting demand.

Many of the lighting technologies used today are considered inefficient in terms of energy use. Incandescent lamps are commonly found in North American and European households, but about 90 percent of the energy used by these forms of lighting is lost as heat. It is estimated that policies to introduce more efficient lighting, like compact fluorescent lamps (CFLs), have saved almost 8 percent of cumulative lighting electricity consumption since the 1990s (IEA 2006, 28). The European Union's recent legislation to ban sales of incandescent bulbs is likely to improve lighting efficiency in member countries, but there are still many uncertainties about the further deployment and use of alternative technologies such as CFLs. For example, CFLs have a much longer life span than incandescent lights but also come at a much greater cost (currently three to ten times higher), and many households indicate that they still prefer the characteristics of light emitted by incandescent bulbs (Mills and Schleich 2010, 376).

Demand for Artificial Lighting

It is only relatively recently that human dependence on artificial lighting has developed. Prior to the advent of electric lighting the source of light for many was from the hearth, supplemented by oil lamps or candles. This changed with the arrival of gas lighting in the early nineteenth century, which was installed in factories, shops, and on the streets, permitting a social life after dark for urban dwellers. The development of electric lighting later that century further transformed the urban landscape, as exemplified in the dazzling electric signage of Broadway's Great White Way (Nye 1990).

Demand for electric lighting did not arise seamlessly or autonomously. When Thomas Edison introduced the incandescent light bulb in 1879, well-established gas companies dominated the lighting of homes and offices. While many Americans initially disliked the brilliance of electric light, preferring the softer glow of gaslight, improvements in fixtures and shading helped to win over consumers. Initially a luxury for the rich minority, by the 1930s electric lighting was becoming not only more affordable but more fashionable within the home. A clear advantage of this new form of lighting was the flexibility in daily routines that it supported; for the first time work could be spread over the day rather than being confined to daylight hours.

The advent of electric lighting has profoundly influenced society. Rather than working with the natural angle and intensity of daylight, we are now able to work to an artificial timetable imposed by modern work schedules. Technological innovations, such as the electric light, have effectively supported the move toward a global twenty-four-hour society. Most people today spend more than 90 percent of their time indoors, and whatever the time of day or night, artificial lighting is critical to providing visual comfort.

Standards of Indoor Lighting

Human beings are accustomed to significant variations in the level and duration of daylight, but standard lighting practice used in building design defines a much narrower optimal range. Natural outdoor illumination can vary from over 100,000 lux (a standard unit of illuminance related to how much light is spread over a given area) on a sunny day to a few thousand lux on a dark, overcast winter day (Altomonte 2008, 7). External lighting levels are therefore, on average, much higher than the standard requirement for work spaces in North America and Europe, which typically stipulate a maintained illuminance of between 300 and 500 lux, depending on the type of activity (Altomonte 2008, 5).

Good lighting also signifies much more than providing a suitable intensity of illumination. The influence of lighting color (measured on a spectrum in terms of shade of "whiteness" of a light source) can improve subjective well-being, but many existing office environments are designed to create neutral and uniform indoor lighting configurations. Poor lighting conditions caused by factors such as glare and flicker are also associated with distraction or fatigue. The scientific consensus is that healthy lighting for daytime indoor activity is influenced by many more factors than those covered by existing lighting standards and regulations.

Heavy reliance on artificial lighting is also associated with daylight deprivation and depression. Concern over the health effects of prolonged working hours in artificially illuminated spaces has led to a debate about the fundamental necessity of natural light. From a historic point of view, the importance of daylight has been reflected in legislation; for example, Roman laws established solar rights for citizens to guarantee access to daylight, while planning principles adopted in Boston and New York at the beginning of the twentieth century banned the dark street canyons emerging in urban centers (Altomonte 2008, 3). Today, however, many current building-design standards contain no absolute legal requirement or numerical guidelines for daylight in the workplace.

Global Trends and Cultural Variations

Aside from direct health concerns, artificial lighting has a high penalty in electricity consumption and overall environmental impact owing to the high level of carbon dioxide emissions of electricity generation. Current lighting practices are often enormously wasteful, with spaces illuminated even when unoccupied, or routine utilization of diffuse sources in buildings rather than applications focused on getting light only where it is needed. Uninspired building designs further accentuate this problem by reinforcing the need for artificial lighting over natural daylighting. But patterns of lighting demand and associated practices also vary enormously around the world. Several recent studies have provided a detailed understanding of lighting practices in different countries and cultures that can help to understand the social context for the implementation of more efficient and sustainable lighting systems.

In some parts of the world lighting is still largely provided by the natural movements of the sun, or by localized light provided from candles and other fuels. At present 1.6 billion people worldwide are estimated to live without access to electric light, relying instead on paraffin or diesel-fueled lighting. These combined forms of lighting provide only 1 percent of global lighting but are responsible for 20 percent of lighting's carbon dioxide emissions (IEA 2006, 25). While the general global trend converges toward electric lighting, there are significant differences in how these technologies are applied within different settings and hence in levels of lighting efficiency. One cross-cultural study of lighting behavior found that Norwegian households typically use many more light bulbs per living room than those in Japan (Wilhite et al. 1996). In Norway multisource lighting (from a combination of table, reading, and spot lamps) is considered essential for creating a mood of coziness, while in Japan ceiling fixtures and fluorescent lights are deemed necessary for spaces to be considered well lit. In addition, the Japanese were much more diligent in turning lights off when leaving the room or the house. The preference for certain qualities of light and routine reproduction of certain lighting habits clearly has an important geographical and cultural dimension that will shape the success of sustainable lighting policies.

Transforming Lighting Practices

Recent advances have continued to focus lighting developments on greater energy efficiency through technological substitution. Toward the end of the twentieth century, the energy-saving compact fluorescent lamp has become more and more accepted both for domestic and commercial applications. In line with this trend, lighting based upon light-emitting diodes (LEDs) is now starting to emerge. These technologies have excellent energy efficiency and an extremely long operating life but have yet to achieve the brilliance or color quality of other lighting technologies.

While technological substitution has the potential to significantly reduce the energy burden of lighting, uptake to date has been somewhat disappointing. It is estimated that energy-efficient CFLs represent only 4 percent of the global market and 6 percent of the European market (Mills and Schleich 2010, 364). Fluorescent lamp adoption is also considerably lower in the residential sector than in the service sector, with many European households indicating that they still prefer the warm yellow glow emitted by incandescent bulbs. Regulatory measures to ban the incandescent bulb may help to increase the rate at which such technical fixes are deployed, but transforming lighting preferences is another necessary policy component.

In parts of the world where electrification has been slow and sporadic, alternative technological solutions are also being developed. In rural India and China, for example, photovoltaic solar lanterns have been suggested as a cost-effective and energy-efficient lighting solution. Such alternatives have considerable potential in terms of quality of illumination, durability, and versatility of use but still face significant social, economic, and cultural barriers to dissemination.

Building design that more effectively utilizes daylight is another essential component of government policies to reduce lighting-related energy demand. Establishing minimum requirements for daylighting in buildings is one priority; however, the energy consequences of daylighting standards are not straightforward. Increasing window area can help to reduce lighting energy consumption, but associated solar-heat gain can increase the energy demand related to artificial cooling. In this context, building regulations need to adopt an integrated approach that connects visual comfort requirements with wider thermal comfort considerations, for example, the need for adequate shading to reduce glare and heat from the sun. Transforming the routine practice of illuminating unoccupied buildings is another significant challenge. In this regard, new legislation that mandates the use of lighting controls to shut off lighting automatically when it is not needed will be an important component of efficient building design.

Other studies suggest that technological solutions to the worldwide lighting energy crisis will have limited application if they do not address cultural preferences or seek to transform lighting habits. Aside from changes to building lighting standards and controls, further savings might be induced through changing patterns of occupant activity to both better utilize natural variations in light throughout the day or seasonally. More fundamentally, there is a need to improve our understanding of human preferences for different forms of lighting to be able to comment on the potential for energy savings from energy efficient applications. Replacing incandescent lamps with compact fluorescent lamps is regarded as a fast and effective measure to decrease household electricity consumption, but understanding culturally formed habits of turning lights on and off and how these are influenced by radically different ideas of visual comfort is just as vital. Another dimension to this debate about lighting cultures is for those setting lighting standards to ensure that less sustainable expectations of lighting quality do not take hold. For example, in China recommended light levels for office spaces are currently only 100–200 lux, but there is evidence of increasing adoption of Western standards of 300 lux or more.

Enlightened Futures?

The current policy focus for indoor lighting encourages the diffusion of more efficient lighting technologies to provide the same or better level of lighting service without using more energy. Along with improving technological efficiency, the revaluing of darkness and of working with natural variations in daylight may become vital factors in the battle to transform human dependencies on energy-intensive and unhealthy lighting practices.

Heather CHAPPELLS
Saint Mary's University

See also in the *Berkshire Encyclopedia of Sustainability* Design, Product and Industrial; Heating and Cooling; Solar Energy

Further Reading

Altomonte, Sergio. (2008). Daylight for energy savings and psycho-physiological well-being in sustainable built environments. *Journal of Sustainable Development*, *1*(3), 3–16.

Baker, Nick; Fanchiotti, A.; & Steemers, Koen. (Eds.). (1993). *Daylighting in architecture: A European reference book*. London: Earthscan.

Bowers, Brian. (1998). *Lengthening the day: A history of lighting technology*. Oxford, UK: Oxford University Press.

Crosbie, Tracey, & Guy, Simon. (2008). En-lightening energy use: The co-evolution of household lighting practices. *International Journal of Environmental Technology and Management*, *9*(2–3), 220–235.

Department of Trade and Industry (DTI). (2002). *Energy consumption in the United Kingdom*. London: DTI.

International Energy Agency (IEA). (2006). *Lights labour's lost: Policies for energy-efficient lighting*. Paris: OECD.

Mills, Bradford F., & Schleich, Joachim. (2010). Why don't households see the light? Explaining the diffusion of compact fluorescent lights. *Resource and Energy Economics*, 32, 363–378.

Nye, David. (1990). *Electrifying America: Social meanings of a new technology, 1880–1940*. Cambridge, MA: MIT Press.

US Energy Information Administration (US EIA). (2008). Residential energy consumption survey. Retrieved October 1, 2010, from http://www.eia.doe.gov/ask/electricity_faqs.asp#electricity_lighting

Wilhite, Harold; Nakagami, Hidetoshi; Masuda, Takashi; Yamaga, Yukiko; & Haneda, Hiroshi. (1996). A cross-cultural analysis of household energy use behaviour in Japan and Norway. *Energy Policy*, *24*(9), 795–803.

Lithium

Lithium is the lightest metal and has many manufacturing and chemical uses. Perhaps its most intriguing potential use is powering hybrid or electric automobiles. To meet this expected demand to create a low-emissions future, lithium mining is increasing, and new sources have been discovered. Lithium is available in hard-rock deposits and in brines, each requiring its own method of extraction.

Lithium is the lightest metal and is widely used in glass and ceramics, as well as increasingly in batteries for electronics and electric vehicles. Minor uses include plastics, photographic film processing, air-conditioning systems, and aluminum smelting, as well as some pharmaceuticals (mainly psychiatric drugs). Lithium mining is growing strongly and is anticipated to expand rapidly if the world moves to hybrid electric and/or fully electric vehicles. Lithium is widely found in hard-rock deposits and brines.

Economic Geology of Lithium Production

Lithium (Li) can be found in a wide variety of minerals, the most important being spodumene, petalite, amblygonite, lepidolite, eucryptite, jadarite, and hectorite. Alternately, lithium is abundant in brines, such as salt lakes, geothermal fluids, or oilfield brines. Historically, lithium production has been relatively small and demand supplied by a small number of mines, mainly hard-rock mines, until the 1980s. As of 2010 brines now dominate production. World annual production in 2010 reached about 25,000 tonnes of lithium (Mohr et al 2010; USGS, 2011).

For hard-rock deposits, the principal economic mineral is spodumene, a lithium aluminum silicate mineral ($LiAlSi_2O_6$) with a greenish appearance. Typical ore grades are 1–3 percent (Li_2O), or 0.5–1.5 percent Li, and mining is typically open-cut or underground mining. Ore is beneficiated to a concentrate grade of 5 percent lithium oxide, which is used directly in glass and ceramics manufacture. The Greenbushes mine in Western Australia, which started in 1982, is the world's largest spodumene mine, supplying almost a quarter of global lithium demand, and it remains one of the largest remaining resources in the world. Spodumene can be converted to lithium carbonate for battery manufacture, but this process remains expensive.

Lithium contained in brines is extensive, especially in the high Andes mountains of South America, in China, and in the western United States. Widely varying estimates of brine-based resources have led to controversy over the extent to which lithium resources could support an electric-vehicle future. Lithium extraction generally involves pumping the brines from salt lakes (known as *salars* in South America), evaporation, and chemical treatment to produce lithium carbonate used in lithium-ion batteries. The critical aspects to assess in determining the viability of brine resources and extraction include the lithium concentration (i.e., brine grade), dissolved magnesium-to-lithium ratio, porosity, fluid density, brine depth, and evaporation rate.

The Future

Significant controversy has erupted over the economic viability of lithium resources that could be relied upon to power an electric vehicle and create a low-emissions future. Estimates of availability have ranged from a low of 10 million tonnes to more than 50 million tonnes, depending on the assumptions used for brine resources. Assuming the electric vehicle future does arrive, the demand for lithium would increase dramatically. This anticipation has led to new exploration and the discovery of new hard-rock and brine resources, such as the new Jadar lithium-borate deposit in Serbia (where the new mineral jadarite was first identified). Given the relative youth of the global lithium mining sector in comparison to gold or copper, there are strong prospects for its future. At present, lithium is not widely recycled, but with the anticipated growth in batteries, more systematic recycling will be needed to ensure ongoing supply, in much the same way as lead is presently recycled from lead acid batteries in vehicles.

Gavin M. MUDD
Monash University

Steve H. MOHR
University of Technology, Sydney

See also in the *Berkshire Encyclopedia of Sustainability* Aluminum; Chromium; Coltan; Copper; Electronics—Raw Materials; Gold; Heavy Metals; Iron Ore; Lead; Minerals Scarcity; Mining—Metals; Nickel; Platinum Group Elements; Rare Earth Elements; Recycling; Silver; Thorium; Tin; Titanium; Uranium

Further Reading

Garrett, Donald E. (2004). *Handbook of lithium and natural calcium chloride: Their deposits, processing, uses and properties.* Amsterdam: Elsevier Academic Press.

Hope, David. (22 March, 2011). Hybrid cars & lithium batteries an economic push in S. America. *Online Journal.* Retrieved August 31, 2011, from http://onlinejournal.com/artman/publish/article_7226.shtml

Mohr, Steve H.; Mudd, Gavin M.; & Giurco, Damien. (2010). *Lithium resources: A critical global assessment.* Prepared for CSIRO Minerals Down Under Flagship—Mineral Futures Collaboration Cluster by the Department of Civil Engineering (Monash University) and Institute for Sustainable Futures (University of Technology Sydney).

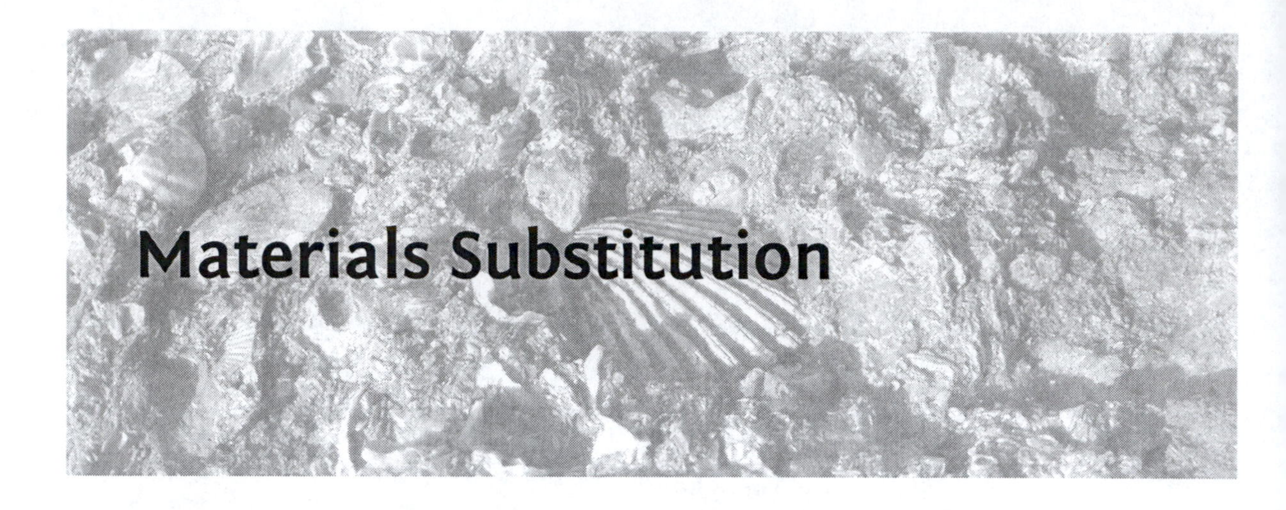

Materials Substitution

Product designers have traditionally chosen materials to optimize a product's form, function, and cost but are increasingly including a product's environmental impacts as a fourth parameter. As designs made with more sustainable material choices gain wider acceptance, companies continue to adopt strategies and tools to assess and reduce the environmental impacts of their materials.

Engineers and designers involved in product development have traditionally made critical choices about materials selection. Although many factors of product design drive those decisions, chief among them have been the product's form, function, and cost. With the consuming public's increasing environmental awareness and demand for "greener" products, and manufacturers' corresponding focus on more sustainable offerings, the environmental impacts of materials have increasingly become a key fourth factor guiding material choice.

Determining Materials Selection

As the first determinant of materials selection, the desired *form* of a product refers to the way that customers, including the end user, will interact with the product. The form comprises a product's overall design aesthetic, including the desired colors, shapes, smells, sounds, and textures, as well as related considerations, such as the ergonomics of use. The product's need to be compatible with or even fit into larger assemblies can also dictate shapes and sizes, fastening mechanisms, and material types.

The *function* of the product, the second key material determinant, refers to those attributes that allow the product to perform its core purpose or purposes. The form of the product could itself fulfill this purpose or be supplemented with features leveraged by mechanical, chemical, electrical, or other means. For example, the form of a baby's toy consists of materials and components that provide pleasing sights, sounds, and textures for the infant. Its function is to safely entertain the infant; in addition to providing pleasant form factors, the materials and shapes must be safe when handled and chewed, and must comply with any regulations governing children's toys.

Designers of a more complex product, such as a mobile phone, must take into account other form and functional factors that constrain the range of available materials. For instance, materials on the printed circuit board of the mobile phone must minimize electrical resistance to transmit power and signal without excessive heat gain; materials that are structural must have adequate strength while maintaining a low weight and small form; and the adhesives used to bond these materials must be able to withstand the physical impact of normal use.

Cost is the third chief determinant of material selection. Designers and engineers often must hit a target cost of goods, or total unit cost for the manufacture and distribution of each product. The simplest contributor to this cost of goods is the direct material cost; for example, silver is superior to copper for bulk electrical conductivity, but it is used sparingly because of its significantly higher unit cost (Davis 2001). Material choice also influences indirect costs, such as the types of manufacturing processes that can be used to engineer the desired form. In some cases, the material selected must be extracted, sourced, grown, refined, or worked in a particular region of the world, which can influence transportation and processing costs as well as labor, energy, infrastructure, and other regionally dependent variable costs.

Finally, engineers and designers consider the *environmental impacts* of their material choices as a fourth determinant of material selection. As society's understanding of environmental consequences evolves, so has the realization that it is no longer enough for products to be inexpensive, functional, and aesthetically pleasing to the consumer. Many products throughout history—leaded paint, asbestos siding, brominated flame retardants (BFRs) in electronics, methyl tert-butyl ether (MTBE) as a fuel additive—have satisfied form, function, and cost specifications, but at a detriment to human and ecosystem health. These environmental effects have led to increases in societal problems such as incidences of sickness and contamination of industrial sites, and ultimately to overall higher financial costs to society.

Societies and governments, increasingly made to bear the burdens of these *externalities* (unintended financial costs that the producers of the materials do not bear, such as health-care increases), are beginning to incorporate the externalities into material and production costs by instituting concepts such as product environmental regulations and extended producer responsibility (EPR). At the same time, leading companies are creating competitive advantages by proactively reducing their products' environmental impacts through methods such as lower-impact material choices. To influence consumer choice, some of these companies are communicating the environmental benefits of their environmental product decisions to consumers, a practice known as *green marketing*.

A material's *environmental footprint*, or its aggregate environmental impacts, is less well understood than the other three material choice determinants. The US Environmental Protection Agency (EPA) was established in 1970 over concerns about hazardous materials, such as the unchecked use of the insecticide DDT, an alarm first sounded by Rachel Carson in her 1962 book *Silent Spring* (Lewis 1985). But the EPA proved not to be a comprehensive solution; the EPA approves about 90 percent of new compounds without restrictions, and only 25 percent of the 82,000 chemicals in use in the United States have been tested for environmental hazard or human toxicity (Duncan 2006). In the European Union (EU), a 2006 regulation known as Registration, Evaluation, and Authorisation of Chemicals (REACH) was enacted to combat this specific problem: all existing and new materials developed in the EU must be tested for environmental hazard, with the burden of proof on the material manufacturer to prove that the material is safe, rather than on the government entity to prove that it is unsafe. In addition, further EU legislation known as Restriction of Hazardous Substances (RoHS), also enacted in 2006, aims to ban specific materials used in electronics products that are known environmental toxins. But global environmental regulation has progressed more slowly than innovation in materials.

Product Life Cycle Approaches

Increasingly, firms are using *product life cycle* approaches to better understand product environmental impacts, such as the comprehensive methodology of life cycle assessment (LCA) governed by the ISO 14040/14044 standards developed by the International Organization for Standardization. While material selection is only one part of a product's life cycle, it affects all stages of the life cycle. As with the product's monetary costs, the choice of materials influences the product's overall environmental burdens through the direct impacts of material extraction and transport, and through indirect effects stemming from the methods and locations of manufacturing. The selection of materials also determines the types of joining technologies (e.g., glues, welds, fasteners) that can be used in product assembly, and the type of packaging that must be employed to protect and transport the product, both of which have environmental consequences. The durability of the chosen materials determines longevity in products that wear out (as opposed to becoming obsolete). Finally, the materials govern the product's end-of-life impacts, determining which components can be recycled, and further determining the environmental effects of incinerating or dumping the remaining components into landfills.

Since material choice has such far-reaching influences on the product's life-cycle environmental impacts, designers and engineers have been working to find lower-impact materials that also satisfy the form, function, and cost requirements of a product. For example, the Samsung/Sprint Reclaim phone uses post-consumer-recycled and plant-based plastic polymers in place of conventional petroleum-based polymers wherever possible. Preserve Products has taken this a step further and manufactures household consumables out of 100 percent recycled materials. Since metals can often be recycled without any degradation of their material properties, many high-volume, single-use metal products incorporate a significant amount of recycled content; for example, aluminum beverage containers in the United States were manufactured with an average 68 percent recycled aluminum content in 2007 (PE Americas 2010). Avery Dennison produces an alternative line of binders made of polypropylene, which is stronger, more easily recyclable, and causes fewer impacts during production than the conventional polyvinyl chloride (PVC). Staples produces another alternative-material office product, a line of notebooks and paper made from bagasse fiber (a waste by-product from sugarcane production) and water- and vegetable-based inks.

These products reflect several broad strategies for selecting materials that perform well on all four key requirements of form, function, cost, and sustainability: the use of post-consumer and post-industrially recycled and further recyclable materials; the use of naturally sourced materials; the creative reuse of waste by-products; and the in-class substitution of materials, such as choosing plastics or metals that result in lower life-cycle environmental impacts than the traditional plastic or metal material choices.

Public Awareness and Supporting Tools

The consuming public is increasingly becoming aware of these strategies and is now looking for product indications that these strategies have been employed. For example, many products bear an indication that a product was made with a certain percentage of post-consumer recycled material, or that the product or its packaging can itself be recycled, in response to consumer demands for reduced material impacts. Disclosed material and chemical lists satisfy public demand for increased material transparency, such as those listed on "green" cleaning agents like Seventh Generation brand. In some cases, so-called ecolabels, such as the Forest Stewardship Council (FSC) certification for sustainably harvested pulp and lumber products, create credibility for these green-marketing campaigns.

To be certain, strategies designed to reduce environmental impacts by lowering material impacts in *production* are only one side of the solution; consumer behavioral changes addressing the *consumption* side of the economic engine are necessary as well. As consumers become increasingly aware of the environmental and societal effects of materials, they can drive innovation and change from their purchasing behaviors. For example, in 2010 the National Toxicology Program and the Food & Drug Administration (FDA) in the United States have raised concerns over the potentially harmful effects on human health and the environment of bisphenol A (BPA), an industrial chemical used in plastics production (Zeratsky 2010). Although these harmful effects have not been conclusively proven and are disputed by plastics industry associations (Bisphenol A 2011), the public furor has caused consumers to seek material alternatives, and manufacturers have begun to provide BPA-free alternatives in response to the consumer demands.

Supporting tools are required in production to help companies achieve these sustainable design strategies and satisfy consumer demands. Design tools are evolving to allow engineers and designers to evaluate the four key material considerations of form, function, cost, and sustainability in the product design process to create successful products. Various software packages allow a product's form to be modeled, function to be simulated, cost to be estimated, and environmental impact to be assessed; several packages allow two or three of these functions to be performed simultaneously and iteratively. Tools that will allow the designer to optimize material choice on all four key criteria are currently under development and will lead to more intelligent material choices in product design and more sustainable resource utilization throughout the world.

Asheen A. PHANSEY
Dassault Systèmes SolidWorks Corp.; Babson College

See also in the *Berkshire Encyclopedia of Sustainability* Design, Product and Industrial; Electronics—Raw Materials; Indigenous and Traditional Resource Management; Industrial Ecology; Minerals Scarcity; Mining—Metals; Mining—Nonmetals; Nanotechnology; Rare Earth Elements; Recycling; Waste Management

Further Reading

Bisphenol A. (2003–2011). Human health & safety: Bisphenol A and consumer safety. Retrieved August 22, 2011, from http://www.bisphenol-a.org/human/consafety.html

Davis, Joseph R. (Ed.). (2001). *ASM specialty handbook: Copper and copper alloys*. Materials Park, OH: ASM International.

Duncan, David Ewing. (2006, October). The pollution within. *National Geographic*. Retrieved November 10, 2010, from http://ngm.nationalgeographic.com/2006/10/toxic-people/duncan-text

European Commission Environment Directorate General. (2007, October). REACH in brief. Retrieved November 10, 2010, from http://ec.europa.eu/environment/chemicals/reach/pdf/2007_02_reach_in_brief.pdf

Lewis, Jack. (1985, November). The birth of EPA. *EPA Journal*. Retrieved November 10, 2010, from http://www.epa.gov/history/topics/epa/15c.htm

PE Americas. (2010, May 21). Life cycle impact assessment of aluminum beverage cans. Retrieved November 24, 2010, from http://www.container-recycling.org/assets/pdfs/aluminum/LCA-2010-AluminumAssoc.pdf

Zeratsky, Katherine. (2010). What is BPA, and what are the concerns about BPA? Retrieved August 22, 2011 from http://www.mayoclinic.com/health/bpa/AN01955

Mining

Mining for metals and other resources from any mine cannot actually be sustainable but there are measures a mining operation can take to improve its environmental and social performance: reduce use of resources (including water) and the production of by-products and residues; design mines for energy efficiency; improve technologies; and eliminate health and safety hazards. Controlling the use of "conflict" minerals that fund civil wars is essential.

Many people are uncomfortable using the words *mining* and *sustainability* together in the same sentence. Mining occurs at locations in the Earth's crust where there are concentrations of specific minerals or metals. The resources that can be economically extracted at these locations by a mine, referred to as *ore*, are finite and therefore not sustainable. A term such as *sustainability mining*, although widely used by academics and others to emphasize the importance of this concept in mining, is clearly not accurate, and some may consider it an oxymoron.

In order to continue the ongoing development of world economies, a sustainable materials supply is essential. A more appropriate indicator than supply alone is the services provided by these materials or their substitutes; for example, copper is an excellent conductor of electricity, but if copper can be replaced by another material and provide the same conductive service at an economically, environmentally, and socially sustainable fashion, then that will occur. At present a large number of these services are being provided by materials extracted from the Earth's crust. Recycling is an important component of the overall materials (or service) supply, however, less than 30 percent of the current supply contains recycled materials, apart from some metal alloys that can be as high as 70 percent recycled (Ashby 2009). Primary extraction of minerals and metals from the Earth's crust will therefore be an ongoing necessity for a sustainable society, and mining will remain an important activity in those areas where there are concentrations of minerals and metals that can be economically recovered within the environmental and social expectations of stakeholders in the surrounding areas.

During the last decade it has become clear that the correct way to express the importance of sustainability concepts in mining is to refer to it as "the contributions that mining makes to sustainability and sustainable development" (MMSD and IISD 2002). At the global level, mining contributes to sustainability and sustainable development through the supply of materials that provide important services to individuals and nations. The contributions that a specific mine makes to sustainability and sustainable development must be addressed at the local, regional, and often national scale. The essential components of mining and sustainability are environment, economics, community, governance, and technology.

A large modern mine is a vast industrial complex that depends on sophisticated technology for the mining process and recovery of the metals in the ore as well as the management of the liquid and solid residual and waste materials. Computer technology and continuous monitoring of the process stream are widely implemented.

Ecological Aspects of Mining

When left unchecked, mining activities have the potential for widespread damage to the surrounding ecosystems. The extraction of a material from the Earth requires energy, water, and the displacement of large quantities of surrounding material. The process of separating a mineral or metal from ore requires the use of many reagents, some of which are hazardous, such as cyanide. For example in one estimate, to obtain a kilogram (kg) of pure gold, a mining company consumes an average of 143 gigajoules of energy; 691,000 liters of water; and 141 kg of cyanide, and releases 10.4 metric tons of carbon dioxide into the atmosphere (Mudd 2008). These requirements increase as the purity (or grade) of the ore decreases; as the highest grade (purest) resources are depleted, the environmental impacts of mining operations have the potential to increase. Surface mining requires the removal of vegetation, topsoil, and overburden, as well as the construction of roads for heavy vehicle traffic. These activities change the local ecosystem and can contribute to erosion. Underground mining can cause the surrounding land to sink, an effect known as surface subsidence. In addition, runoff from waste rock and tailings (material left over after the metals or minerals have been removed) can leach and transport metals, acid, residual reagents such as cyanide, and sediment into nearby waterways. In some cases the runoff from mining operations can adversely affect the surrounding fish, aquatic organisms, plants, and human populations. Such contamination can continue for years after a mine has closed. Blasting, crushing rock, and storing and transporting materials may all release dusts (potentially containing heavy metals) into the surrounding atmosphere, where winds can carry them miles from the mine.

Much of the ecological damage caused by mining can be prevented or repaired through careful planning, adequate safeguards, careful decommissioning of the mine, and modern reclamation techniques. Ecologically sound practices increase the costs of operations. In developed countries, such as the United States, less-destructive practices and financial assurance for reclamation are required before a mining permit is issued, and ecological effects are monitored by regulatory agencies before, during, and after mining. Mining companies are expected to internalize the costs of protecting and reclaiming the mine site. In less developed areas, the regulations may be of the same standard but the capacity to implement and enforce them is poorly developed. The ecological failures of mining operations may affect a company's future access to mineral resources, even in countries with more relaxed regulations. In 2003 for example, residents of a small community in Argentina decided in a local referendum not to allow Meridian Gold to develop a mine in the region (Turner 2005). This resulted in a significant drop in the share price of the company. Similarly, Vedanta Resources has had investors sell stakes amid concerns over its bauxite-aluminum operations in India's Orissa province; villagers in the area have been fighting the expansion of the project, which they see as a threat to the forest that sustains their way of life (Jena 2010).

Several measures to reduce the ecological impacts of mining are already in use in many places (Spitz and Trudinger 2009). Reusing water and dewatering tailings (or residues) can reduce the consumption of this resource, which can be scarce in arid and semiarid climates of South America and Africa. Reducing the moisture content in tailings to allow deposition as a paste or drier filter cake will reduce the infiltration of contaminated waters into the subsoil. Unfortunately this measure is more costly than pumping the tailings as a slurry into a tailings management facility. These facilities may leak, and there have also been a number of tailings impoundment failures over the years with severe human casualties and impacts to the environment.

Waste rock can be covered to prevent the infiltration of water that can lead to metal and other chemical constituent releases. Tailings and waste rock can also be used to backfill areas of a mine and in some cases are mixed with cement to provide further stability in underground mines.

Designing a mine for energy efficiency can reduce fuel and electricity consumption and operating costs; in many cases, the proper sizing of pumps, motors, and pipes can save energy (Southwest Energy Efficiency Project 2010). Careful planning and the use of dust control equipment can abate the ecological impact of blasting, ore processing, and transportation. Mine closure activities, including reclamation that is meant to provide for successful future land use such as grazing, must be planned from the initial stages of mine design and development.

Some promising improvements to mining technology and methods can lower operating costs while decreasing the ecological risks of the operations. By using the most advanced prospecting and exploration technology available, such as remote sensing, mining companies can better pinpoint the locations of deposits, which would allow them to minimize the amount of surface disturbance during exploration activities. Biologically based processing techniques have the potential to eliminate the use of some hazardous chemicals while improving recovery rates.

Socioeconomic Aspects of Mining

While mining operations have the potential to bring material wealth to the surrounding communities, they can also bring social and economic harm. Without adequate safety measures, mines—especially those below ground—pose many hazards to workers. While the number of mining-related fatalities reduced dramatically in the United States over the last century, this is not the case internationally. In 2002, nearly 7,000 miners died in China's coal mines (China Labour Bulletin 2006). Because mining requires expensive equipment and/or many laborers, unethical mine owners have been known to exploit workers. When armed security forces are used to protect mining interests, human rights violations may occur. Profits from mining can be used to finance armed conflicts. In eastern areas of the Democratic Republic of Congo, for instance, militant groups thrive on the mining of coltan (a source of tantalum, an essential component of the capacitors used in almost every electronic device). These armed men take a portion of everything that comes from the mines and commit violence against the local populace, using funds from the illicit sale of coltan and other "conflict minerals" to continue a war with the government (Allen 2009).

Mining operations should generate revenue and jobs for host countries that should benefit the local populations. The redistribution of mine taxes to local authorities is one mechanism for accomplishing this. These mechanisms, however, may not be well-developed in a host country, and therefore most of the wealth doesn't reach the communities surrounding the mines. A 2007 study of communities in Ghana indicated that towns closer to mining projects were generally poorer than their more distant counterparts (Akabzaa 2009). Large-scale surface mining operations rely on heavy equipment for the work, meaning fewer jobs are created than in small-scale mining, and those jobs often require skills that the local workers do not have. In Ghana, for example, only 0.7 percent of the working-age population is employed by large-scale mining companies (Akabzaa 2009). Mining interests can impact other resources (i.e., land) and limit their use by other industries, such as farming.

Artisanal mines, which generally are small-scale local or regional operations, can involve thousands of local workers on a subsistence economic level. On a global scale it is estimated that tens of millions of workers are involved with such activities in the recovery of gold, diamonds, gems, and other minerals. These operations rely on simple and often dangerous practices since they are not licensed and regulated. In gold mining, mercury amalgamation is used by many, and it results in widespread human health and environmental impacts. The relations between artisanal mining and large-scale mining are often tense because artisanal miners may occupy the land that a large mining company hopes to access. The ore grades of most large mines are much lower than most areas mined by artisanal miners, albeit these smaller areas can not be economically developed by large companies.

Mining does not have to exact a high social and economic cost. In fact, by working with community members to ensure their needs are met, a mining operation can be good for the community. Colorado-based Newmont Mining Company, for example, has implemented a program for ensuring good relationships with the communities in which it works in North and South America, Africa, and the Asia/Pacific region. The program involves interviewing community members, workers, and other stakeholders to evaluate and improve the company's social impact. The program led the company to help farmers in fifteen Indonesian villages around copper and gold mines to diversify their crops and improve their livelihoods (Newmont 2009).

Legacy Issues

Throughout the world there are many environmental and social legacy issues associated with mining. These legacy sites, also referred to as abandoned and orphaned mines, were previously operated but not properly closed because the regulatory framework did not exist. The major focus of present regulatory processes and other activities is to make sure that no more legacy sites are created. For example, in the United States alone it is estimated that 19,300 kilometers of rivers and streams and more than 730 square kilometers of lakes and ponds have been adversely affected by acid drainage from abandoned mines (Montana State University 2004).

Environmental cleanup at abandoned mine sites is typically paid for by a government, and it is very dependent on national economic conditions. The first priority for abandoned mine sites is to make them safe for the public that may wander into the area. A small percentage of the abandoned sites require extensive environmental remediation, which is usually very costly.

A multistakeholder process in Canada has been very successful in developing strategies and engagement processes on the issue of abandoned mines. The National Orphaned/Abandoned Mines Initiative (NOAMI 2009) has conducted a series of workshops and published a series of reports on these topics.

A Framework for Assessing Sustainability

At the global level, mining contributes to sustainability and sustainable development through the supply of materials that provide important services to individuals and nations. In 2001 a group of stakeholders consisting of academics, representatives of nongovernmental organizations, and mining industry representatives and regulators participated in a series of workshops and writing sessions in North America as part of the Mining, Minerals and Sustainable Development Project; the purpose was to establish the means to evaluate the contributions that a specific mining venture—existing or proposed—makes to sustainability during its design, operation, closure, and postclosure. The objective would also suggest approaches or strategies for effectively implementing such a test/guideline (MMSD and IISD 2002). The project's outcome, referred to as the "seven questions to sustainability," is currently used by researchers and practitioners alike in various assessments (Hodge 2004; Van Zyl, Lohry, and Reid 2007). The seven topics or themes "questioned" are engagement, people, environment, economy, traditional and nonmarket activities, institutional arrangements and governance, and synthesis and continuous learning. (See table 1.)

Availability of Materials

In *The Limits to Growth* (Meadows et al. 1972), a global nongovernmental organization known as the Club of Rome issued a report on the potential scarcity of materials. While there has been much discussion about this topic and also evidence that the predictions were not substantiated, this theme is currently receiving renewed attention (e.g., Shields and Šolar, forthcoming). Generally, two models are put forth in the evaluation of this issue. The "fixed stock" model approaches the availability of materials from the Earth as finite and assumes that the demand will eventually exhaust the available supply. The "opportunity cost" paradigm assesses resource availability by what society has to give up to produce another unit of a mineral commodity, such as a tonne of copper. Over time, depletion increases the opportunity cost of mineral production, while new technology and other factors can offset this pressure (Tilton 2003). Improvements in ore processing, for instance, can turn unprofitable deposits into viable resources. Further advances in the recycling of materials may also alleviate some of the demand. As the supply of a material decreases, economic pressure is likely to prompt the search for an alternate material. In short, the sudden and total depletion of a mineral resource is not likely to occur.

Table 1. **Seven Questions to Sustainability**

Does the project identify stakeholders and engage them in all phases, from planning to closure?
Will peoples' well-being be maintained or improved during and after the project or operation?
Will the integrity or well-being of the environment be maintained or improved during and after the project or operation?
Is the economic viability of the company assured; is the community and regional economy better off not only during operation but into postclosure?
Is the viability of traditional and nonmarket activities in the community and surrounding area maintained or improved with the project or operation?
Are the rules, incentives, and capacities in place now and as long as required to address project or operational consequences?
Does a synthesis show the project to be net positive or negative for people and ecosystems; is the system in place to repeat the assessment from time to time?

Source: (Hodge 2004; Van Zyl, Lohry, and Reid 2007)

Researchers, industry professionals, and other stakeholders have established seven questions to consider in assessing the environmental and socioeconomic impacts of a mining activity.

Outlook for the Twenty-First Century

Pressure from shareholders, communities, governments, and materials buyers will push mining companies toward improved performance. The Church of England and other investors sold their shares in the global metals and mining group Vedanta Resources in 2010 due to publicity surrounding the company's operations in India, and in Guatemala, people in forty-two municipalities voted to prohibit mining. In 2007, labor groups in Guinea staged an uprising that forced the government to reevaluate its mining contracts (Campbell 2009). Organizations such as Oxfam America have been working to protect the rights of communities near mines. Major retailers including Walmart and Tiffany's have pledged to avoid gold suppliers guilty of human rights or environmental violations.

Already some of the largest mining companies are making changes to the way they conduct business in order to ensure their futures. In 2001 the International Council on Mining and Metals (ICMM) was formed "to act as a catalyst for performance improvement in the mining and metals industry. Today, the organization brings together 19 mining and metals companies as well as 30 national and regional mining associations and global commodity associations to address the core sustainable development challenges faced by the industry" (ICMM 2005). The ICMM sustainability framework consists of three parts: a set of ten principles, reporting on their performance, and assurance of the reports. ICMM also publishes a large number of guidance documents (such as the *Community Development Toolkit*) that are widely used by various stakeholders.

Growing awareness of the contributions that mining can make to sustainable development should prompt many mining companies toward improved environmental and social responsibility. While there are clear leaders in the industry, the overall performance of the industry is often judged by that of the laggards.

Dirk VAN ZYL
Norman B. Keevil Institute of Mining Engineering, University of British Columbia, Vancouver

David GAGNE
Berkshire Publishing Group

See also in the *Berkshire Encyclopedia of Sustainability* Base of the Pyramid; Cement Industry; Development, Rural—Developing World; Energy Efficiency; Energy Industries—Coal; Human Rights; Investment, CleanTech; Poverty; Steel Industry; Telecommunications Industry; Water Use and Rights

Further Reading

Akabzaa, Thomas. (2009). Mining in Ghana: Implications for national economic development and poverty reduction. In Bonnie Campbell (Ed.), *Mining in Africa*. Retrieved February 24, 2010, from http://www.idrc.ca/en/ev-141150-201-1-DO_TOPIC.html

Allen, Karen. (2009, September 2). Human cost of mining in DR Congo. Retrieved March 1, 2010, from http://news.bbc.co.uk/2/hi/8234583.stm

Aryeetey, Ernest; Bafour, Osei; & Twerefou, Daniel Kwabena. (2004). *Globalization, employment and livelihoods in the mining sector of Ghana* (ISSER Occasional Paper). Accra: University of Ghana.

Ashby, Michael F. (2009). *Materials and the environment: Eco-informed material choice*. Oxford, UK: Butterworth-Heinemann.

Campbell, Bonnie. (2009). Guinea and bauxite-aluminum: The challenges of development and poverty reduction. In Bonnie Campbell (Ed.), *Mining in Africa*. Retrieved February 24, 2010, from http://www.idrc.ca/en/ev-141151-201-1-DO_TOPIC.html

China Daily. (2007, November 30). Blueprint for coal sector. Retrieved January 18, 2010, from http://www.china.org.cn/english/environment/233937.htm

China Labour Bulletin. (2006). Deconstructing deadly details from China's coal mining statistics. Retrieved February 24, 2010, from http://www.clb.org.hk/en/node/19316

Engels, J., & Dixon-Hardy, D. (2010). Tailings.info. Retrieved February 24, 2010, from http://www.tailings.info/index.htm

Gordon, R. B.; Bertram, M.; & Graedel, T. E. (2006). Metal stocks and sustainability. *Proceedings of the National Academy of Sciences*, 103(5), 1209–1214.

Hodge, R. Anthony. (2004). Mining's seven questions to sustainability: From mitigating impacts to encouraging contributions. *Episodes*, *27*(3), 177–184.

International Council on Mining and Metals (ICMM). (2005). Community development toolkit. Retrieved March 10, 2010, from http://www.icmm.com/page/629/community-development-toolkit-

Jena, Manipadma. (2010, February 23). India: Indigenous groups step up protests over mining projects. Retrieved February 26, 2010, from http://ipsnews.net/news.asp?idnews=50429

Li, Ling. (2007). China's largest coal province launches sustainable mining fund. Retrieved February 24, 2010, from http://www.worldwatch.org/node/4992

Meadows, Donella H.; Meadows, Dennis L.; Randers, Jorgen; & Behrens, W. W. (1972). *The limits to growth*. New York: Universe Books.

Millennium Ecosystem Assessment. (2005). *Ecosystems and human well-being: Current state and trends, Vol. 1*. Washington, DC: Island Press.

Mining, Minerals and Sustainable Development. (2002). *Breaking new ground: The MMSD final report*. Retrieved October 1, 2009, from http://www.iied.org/pubs/pdfs/9084IIED.pdf

Mining Minerals and Sustainable Development North America (MMSD) & International Institute for Sustainable Development (IISD). (2002). *Seven questions to sustainability: How to assess the contributions of mining and minerals activities*. Winnipeg, Manitoba: IISD.

Montana State University, Bozeman. (2004). Environmental impacts of mining: Acid mine drainage formation. Retrieved February 24, 2010, from http://ecorestoration.montana.edu/mineland/guide/problem/impacts/amd_formation.htm

Mudd, Gavin. (2008). Gold mining and sustainability: A critical reflection. In *The encyclopedia of Earth*. Retrieved February 24, 2010, from http://www.eoearth.org/article/Gold_mining_and_sustainability~_A_critical_reflection

National Mining Association. (1998). *The future begins with mining: A vision of the mining industry of the future*. Retrieved February 26, 2010, from http://campus.mst.edu/iac/iof/industies/MINING/mining_vision.pdf

National Orphaned/Abandoned Mines Initiative (NOAMI). (2009). *Performance report 2002–2008*. Retrieved March 10, 2010, from http://www.abandoned-mines.org/pdfs/NOAMIPerformanceReport2002-2008-e.pdf

Newmont Mining Company. (2009). Batu Hijau, Indonesia: Helping farmers transition from subsistence to surplus. Retrieved February 27, 2010, from http://www.newmont.com/asia-pacific/batu-hijau-indonesia/community/farmers-subsistance-surplus

Rajaram, Vasudevan; Dutta, Subijoy; & Parameswaran, Krishna. (2005). *Sustainable mining practices: A global perspective*. London: Taylor & Francis.

Shields, D., & Šolar, S. (forthcoming). Responses to alternative forms of mineral scarcity: Conflict and cooperation. In S. Dinar (Ed.), *Reflections on resource scarcity and degradation: Conflict, cooperation and the environment*. Cambridge, MA: MIT Press.

Southwest Energy Efficiency Project. (2010). Energy efficiency guide for Colorado business, recommendations by sector: Mining. Retrieved February 26, 2010, from http://www.coloradoefficiencyguide.com/recommendations/mining.htm

Spitz, Karlheinz, & Trudinger, John. (Eds.). (2009). *Mining and the environment: From ore to metal*. London: Taylor & Francis Group.

Tilton, John E. (2003). *On borrowed time?: Assessing the threat of mineral depletion*. Washington, DC: Resources for the Future Press.

Tribal Energy and Environmental Energy Clearinghouse. (n.d.). Coal mining: Decommissioning and site reclamation impacts. Retrieved February 27, 2010, from

Turner, Taos. (2005). South America mining industry sees investment boom. Retrieved February 26, 2010, from http://www.minesandcommunities.org/article.php?a=485

Van Zyl, D.; Lohry, Jerome; & Reid, R. (2007). Evaluation of resource management plans in Nevada using seven questions to sustainability. In Z. Agioutantis (Ed.), *Proceedings of the 3rd International Conference on Sustainable Development Indicators in the Mineral Industries* (pp. 403–410). Milos Island, Greece: Milos Conference Center-George Eliopoulos.

Petroleum

We live in the age of petroleum. Oil and natural gas provide two-thirds of our energy as well as feedstocks for most chemicals, plastics, and other accoutrements of modern life. Historically there has been a close correlation between petroleum use and economic activity, and the recent faltering in global energy availability (peak oil—perhaps) has been attributed as the critical determinant of many financial problems.

The word *petroleum* usually refers to natural liquid and gaseous hydrocarbons, and includes oil, natural gas, and natural gas liquids. Sometimes the term is used to describe oil alone. These substances, along with coal and certain other low-quality solids, are considered *fossil fuels*. Gas is often found associated with oil, although it has other possible sources, including coal beds and organic-rich shale. What we call oil is actually a large family of diverse hydrocarbons whose physical and chemical qualities reflect their different origins and, especially, different degrees of natural processing.

Petroleum is an extraordinarily useful resource from which humans derive fuels, plastics, asphalt, nitrogen fertilizers, tires, paint, and a whole host of organic chemicals ranging from aspirin to xylene. Petroleum possesses unique qualities, including high energy density and transportability. But due to the magnitude of its use in the twentieth and twenty-first centuries, its future supply has become a cause for concern that requires us to examine the relation between potential demand and possible supply.

Origins and Hydrocarbon Comparisons

The origins of most oil and gas occurred during two principal geological times (90 and 150 million years ago). Small freshwater or marine plants (phytoplankton) grew profusely in conditions where they sank into anaerobic (oxygen-free) basins, such as deep rift lakes. Subsequent heavy rains covered the organic material, which had been protected from oxidation, with sediments that eventually pressure-cooked the biotic material over many tens of millions of years (Tissot and Welt 1978). The original plant material, often comprising hundreds to thousands of carbon atoms linked together, was broken or "cracked" by geological energies to a length of (ideally) eight carbons (octane). If the cracking continued to the extreme, the carbon bonds were broken completely to a length of one carbon, usually surrounded by four molecules of hydrogen (natural gas, also called methane). Oil and gas tend to migrate upward from the source rocks to the reservoirs in which they are found, typically under an impervious "cap rock." The vast proportion of oil and gas ever formed probably escaped to the atmosphere without being captured in an exploitable reservoir. From a geological perspective, oil is thus a rare substance because the many complex conditions required for manufacture and capture were rare.

Petroleum can be broadly categorized as *conventional* and *unconventional*. Conventional petroleum refers to a diversity of liquid and gaseous fuels. These substances are derived from geologic deposits, usually found and exploited using drill-bit technology, that move to the surface because of their own pressure, by pumping, or with additional pressure supplied by injecting natural gas, water, or (occasionally) other substances into the reservoir. What we call oil is actually a diverse suite of materials, and its density can vary from 740 to 1,030 kilograms per cubic meter. Unconventional petroleum includes shale oil, tar sands, and other bitumens usually mined as solids, as well as coal bed, "tight sand," shale, and certain other methane deposits that are extracted as gas—often after special treatment such as rock fracturing.

In theory hydrogen is the most potent fuel, but natural gas is in some ways the ideal fuel: oxidizing hydrogen releases more energy and less carbon dioxide than oxidizing carbon, and methane is much more easily obtained, stored, and moved than is hydrogen. When natural gas is held in a tank, some heavier fractions fall out as natural gas liquids, and these materials can be used either directly or as inputs to refineries.

Carbohydrates are less potent fuels than hydrocarbons because they are already partially oxidized. For example, alcohol (methane) fuel from corn contains only about 70 percent of the energy per liter as gasoline. Natural gas liquids are also less energy dense, and contain roughly three-quarters the energy per volume as oil. The utility of petroleum comes from its ability to combine with oxygen to release energy. The shorter the molecule, the higher the hydrogen-to-carbon ratio and usually the greater the energy density.

Using Petroleum

Petroleum is probably contemporary civilization's most important resource beyond sunshine, clean water, and soil. While the technology of the large-scale production and use of petroleum was developed first and most strongly in the United States and Canada, it has spread throughout the world. Food production, economies, and cultures of nearly all nations are heavily petroleum-dependent. Nearly two-thirds of the energy used to run the world economy and most industrial countries—such as the United States, the United Kingdom, Germany, Japan, and Brazil—is based on conventional petroleum, with about the same proportion of use as in the mid-twentieth century. There is a fairly strong correlation between wealth production and petroleum use for most countries (Hall and Klitgaard 2011). In more heavily developed regions, such as North America, Europe, and eastern Asia, nearly everything humans do is based on cheap oil. Where people live relative to where they work, what they do to make a living, how productive they are in the workplace, how much leisure time they have and how they spend it, the price they pay for food and other commodities, and how much education can be made available—all are largely dependent on adequate supplies of cheap oil. For example, it takes the energy of about a gallon of oil a day to feed each of us, about eighty barrels of oil a day to provide an undergraduate's college education, and the energy equivalent of about ten gallons of oil a day to keep us supplied with all the goods and services we demand through our economic activity. In earlier times this level of energy affluence was available to only an elite sector of society, and was usually provided by slave labor or indentured servitude. Researchers have extrapolated how this labor "translates" in terms of twenty-first century oil consumption, based on the calculation that a person uses about 3,600 kilocalories a day, which is about one-tenth the energy of a gallon of oil, and the fact that a person and a gasoline engine work at about 20 percent efficiency. Comparing the very large quantity of fossil fuels used by highly developed nations to human energy output suggests that each person (in Europe and Japan) has the equivalent of thirty hard-working laborers (and in the United States, the most-oil-consuming country, as many as a hundred), to "hew our wood and haul our water" as well as to grow, transport, and cook our food; make, transport and import our consumer goods; get to work; provide sophisticated medical and health services; visit our relatives; and take vacations in faraway places (Cleveland et al. 1984, Hall et al. 1986, Hall and Klitgaard 2011). To put this into a different perspective, a North American's hot shower in the morning consumes far more energy than used by some two-thirds of the Earth's human population in an entire day (Hall, Powers, and Schoenberg 2008).

Oil is especially important for the transportation of people and their goods and services, and as fuel for heating, cooking, and industry. It is a critical feedstock for fertilizers, plastics, most chemicals, and a vast array of contemporary products. Increasingly, less-developed countries are becoming as dependent upon oil as are fully developed nations, for development and increased petroleum use go hand in hand. Some people say that we live in an information age or a postindustrial age. Many others believe it far more accurate to consider the times we live in as *the age of petroleum*, for petroleum is the foundation of our economy and nearly everything we do.

Quality of Petroleum

Oil is a fantastic fuel, relatively easy to transport and useful for many applications; it is very energy dense, and extractable with relatively low energy cost and (usually) low environmental impact. In general, humans have first exploited the "higher quality," that is to say, larger reservoirs of shorter-chain "light" oil resources, because larger reservoirs are easier to find and access, and lighter oils are more valuable and require less energy to extract and refine. Therefore, over time we turn to "lower quality" resources—meaning that we begin the exploitation of increasingly small, deep, offshore, and heavy resources. Whatever the quality, oil must first be found, then the field developed, and then the oil extracted carefully over a cycle that typically takes decades.

Oil in the ground rarely resembles the thick liquid stored in an oil can; it appears more like an oil-soaked brick. The oil must be pushed slowly by pressure to a collecting well. The rate at which oil can flow through these reservoirs depends principally upon the physical properties of the oil itself and of the geological substrate, and also upon the pressure behind the oil. The gas in the well initially provides this pressure. Then, as the field matures, the pressure necessary to force the oil through the substrate to the collecting wells is supplied increasingly by pumping more gas or water into the structure. Detergents, carbon dioxide, and steam—in a process called EOR (enhanced oil recovery)—have been used since the 1920s to increase yields.

Extracting oil too rapidly can cause compaction of the oil-bearing strata (i.e., reservoir) or fragmentation of flows, both of which reduce yields. So our physical capacity to produce oil depends upon several factors: our ability to keep finding large oil fields in regions that we can reasonably access, our willingness to invest in exploration and development, and our willingness to not produce too quickly. Progressive depletion also means that oil in older fields, which once came to the surface through natural drive mechanisms such as gas pressure, must now be extracted using energy-intensive secondary and enhanced technologies. Technological progress in finding and exploiting oil is thus in a race with the depletion of higher-quality resources.

Another aspect of the quality of an oil resource is that oil reserves are normally defined by their degree of certainty and their ease of extraction, classed as "proven," "probable," "possible," or "speculative." In addition, there are unconventional resources such as heavy oil (e.g., the Kern River field in California and large deposits in Venezuela), deep-water oil (e.g., the Tupi field off Brazil), oil sands in Canada, and shale oils in Colorado that are large but very energy-intensive to exploit. There are large quantities of oil left in the world, but the quality of the actual fields is decreasing as we find and deplete the best ones. Now it is increasingly difficult to find the next field, and—as they tend to be smaller, deeper, offshore and/or of poorer quality—more and more energy is needed to extract and refine the oil to something we can use. A forthcoming issue of the journal *Sustainability* (in press as of September 2011 and edited by Charles Hall) presents evidence that the EROI (energy return on investment) is declining over time. At some point it may take a barrel of oil to find and exploit one barrel of oil. At that point the oil age will be over.

Quantity of Petroleum

Most estimates of the remaining quantity of conventional oil resources are based on the carefully considered opinions of geologists and others familiar with a particular region. The term *ultimate recoverable resource* (URR, often written as EUR) is an estimate of the total quantity of oil that will ever be produced from a field, nation, or the world, including the amount of oil already extracted from it at the time the estimate was made. (See table 1.)

Table 1. Published Estimates of World Oil Ultimate Recovery, in Chronological Order.

Source	Est. Volume*	Source	Est. Volume*
USGS, 2000 (high)	3.9	Nehring, 1978	2.0
USGS, 2000 (mean)	3.0	Nelson, 1977	2.0
USGS, 2000 (low)	2.25	Folinsbee, 1976	1.85
Campbell, 1995	1.85	Adam and Kirby, 1975	2.0
Masters, 1994	2.3	Linden, 1973	2.9
Campbell, 1992	1.7	Moody, 1972	1.9
Bookout, 1989	2.0	Moody, 1970	1.85
Masters, 1987	1.8	Shell, 1968	1.85
Martin, 1984	1.7	Weeks, 1959	2.0
Nehring, 1982	2.9	MacNaughton, 1953	1.0
Halbouty, 1981	2.25	Weeks, 1948	0.6
Meyerhoff, 1979	2.2	Pratt, 1942	0.6

*Volume estimates are in trillions of barrels.

Source: Hall et al. 2003.

Estimates of the amount of recoverable oil remaining in the world have ranged dramatically over the last several decades, from a low of 600 billion barrels of oil to a high of 3.9 trillion barrels of oil; estimates have been somewhat stable since 1959 except for the more optimistic estimates of USGS 2000. These values include the 1.1 trillion barrels we have already extracted.

The larger estimates of the United States Geological Survey in 2000 apparently reflect the opinion of economists who believe that price signals will allow lower grades of oil to be exploited through technical improvements, and that earlier conservative estimates will thus be corrected. The lower values tend to be made by the USGS staff geologists.

Even lower estimates come from several high-profile analysts, including Jean Laherrere, many of them retired petroleum geologists with long histories in the oil industry. They suggest that the URR is no greater than about 2.3 trillion barrels, and may be even less. Other estimates, for example, by US and European energy agencies (EIA and IEA, respectively) are coming in increasingly on the low side in agreement with Laherrere. Analyses of long-term data show that the world is now producing two to four barrels for each barrel found. Logically it seems that the best way to find and produce more oil would be to drill more, but in fact the finding of oil and gas is almost independent of drilling rate, at least at the levels that the United States, for example, has been used to undertaking, because determining the next good place to drill takes time.

Pattern of Use Over Time

The US geologist Marion King Hubbert (1903–1989), who derived the best-known model of oil production, proposed that the discovery and production of petroleum over time would follow a single-peaked, more or less symmetric, bell-shaped curve. A peak in production would occur when 50 percent of the URR had been extracted (he later opined that there may be more than one peak). This hypothesis seems to have been based principally on Hubbert's intuition, and it was not a bad guess: he famously predicted in 1956 that US oil production would peak in 1970, which in fact it did. Hubbert also expected the US production of natural gas to peak in about 1980, which it did, although it has since shown signs of recovery, and there is a second peak in 2010 based on "unconventional" and "shale gas." He estimated that world oil production would peak in about 2000. In fact, oil production continued to increase until 2005, after which it appears to have entered an "undulating plateau," as predicted earlier by geologist Colin Campbell. In the past decade, a number of "neohubbertarians" have made predictions about the timing of peak global production using several variations of Hubbert's approach. Various forecasts of the timing of global peak have ranged from one predicted for 1989 (made in 1989) to many predicted for the first decade of the twenty-first century, and even one as late as 2030 (all of these predictions, and the data, are covered in Hall and Klitgaard 2011). Most of these studies assumed world URR volumes of roughly 2 trillion barrels and that oil production would peak when 50 percent of the ultimate resource had been extracted. In comparison, the USGS low estimate (which the survey states as having a 95 percent probability of being exceeded) is 2.3 trillion barrels.

The US physicist Albert Bartlett fitted the left-hand side of Hubbert-type curves to data on actual production while constraining the total quantity under the curve to 2, 3, and 4 trillion barrels for world URR. The high predictions begin with an a priori assumption about a large volume of ultimately recoverable oil. The resultant peaks were predicted to occur from 2004 to 2030. Adam Brandt (2007), a researcher on the environmental impact of fossil fuels, shows that the Hubbert Curve is a fairly good predictor for most post-peak nations because it has occurred for the great majority of oil-producing nations. Other recent Hubbert-type analyses (Kaufmann and Shiers 2008; Nashawi, Malallah, and Al-Bisharah 2010) suggest peaks in about 2013–2014, consistent with the low URR estimates of Campbell and Laherrere, unless there is much more recoverable oil than seems likely at this time. If that is the case, the peak may be displaced for one or two decades. The peak may also be delayed by the general economic slowdown that has been occurring since 2008.

The first half of the age of oil was one of year-by-year growth. The second half will see the continued importance of oil but show a year-by-year decline in supply, with possibly an "undulating plateau" at the top and some help from still-abundant natural gas separating the two halves and buffering the impact somewhat for a decade or so. Analysts such as Charles Hall believe that it will not be possible to fill the growing gap between supply and demand of conventional oil with liquid biomass alternatives, to cite one example, on the scale required—and even were that possible the investments and time required to do so would mean that progress in that direction should have started by 1990. When the decline in global oil production begins, the ensuing "end of cheap oil" will create a very different economic climate.

The actual data on conventional oil production for the world certainly shows a peak or at least an "undulating plateau," and perhaps even a production peak in 2005. This is astonishing, as it happened during times of increasing oil prices. Why is global oil production decreasing, or at least no longer increasing at 3 percent a year as it did for many decades? The principle reason is that most oil production comes from very large oil fields (so-called elephants), and we have found essentially no elephants since the 1960s. Now these large oil fields are aging, and their production is falling. According to Chris Skrebowski, editor of *Petroleum Review*, about one- quarter of the four hundred largest oil fields in the world are in decline, and it appears impossible that new oil discoveries, most of which are not large, can make up for the decline in the elephants.

Energy Return on Investment

The term *energy return on investment* (EROI or EROEI) refers to the amount of usable energy attained compared to the amount of energy expended to obtain it. When the numerator and denominator are derived in the same units, as they should, it does not matter if the units are barrels of oil per oil barrel or kilocalories per kilocalorie, as the results are expressed as a ratio. The average EROI for the *finding* of US domestic oil has dropped from greater than 1000 kilojoule returned per kilojoule invested in 1919 to about 5:1 in 2011. The EROI for *producing* petroleum in the United States has declined from about 30:1 in the 1970s to about 10:1 in 2011 (Guilford et al. in press). Similar declines in EROI are found in many oil-producing regions around the world including Norway and China. This is a consequence of decreasing energy returns from increasingly depleted oil reservoirs, and of increases in energy costs as exploration and development shift deeper and offshore. The ratio for production mostly reflects pumping out oil fields that are a half-century-or-more old, since few significant new fields are being discovered. The increasing energy cost of a marginal barrel of oil (or gas) is one of the major factors behind its increasing monetary cost.

The EROI for global oil and gas (at least for that which was publically traded) was roughly 23:1 in 1992, increased to about 33:1 in 1999, and since then has fallen to approximately 18:1 in 2005. For both the United States and for the world, the EROI strongly reflects drilling effort, so that increased drilling decreases the EROI (Hall and Cleveland 1981; Guilford and Hall in press). If the rate of decline continues linearly for several decades then it would take as much energy in a barrel of oil to get a new barrel of oil. More probably the decline will not be linear, but the energy cost of a barrel of oil will continue to increase nonetheless. This will have enormous impacts on our economy: the EROI needed to undertake some activities, such as driving a truck, is far more than just what is needed to get the fuel out of the ground, as the authors of "Peak Oil, EROI, Investments and the Economy in an Uncertain Future" explain (Hall, Powers, and Schoenberg 2008). They also acknowledge the minimum EROI that we need to run what we call civilization is uncertain, but it is probably somewhere between 5:1 and 10:1.

Impacts of Peak Oil and Declining EROI

In the eyes of many investigators, modern society and its economies are completely dependent upon petroleum. If peak oil and a serious decline in EROI were to occur, the effects would be enormous. To many, including Charles Hall and a number of his colleagues and co-authors cited in the Further Reading section at the end of this article, the evidence that peak oil has been reached, or will be within years and not decades, is as clear as nearly anything in science can be; additionally the EROI for whatever oil is left to exploit has been declining rapidly if not precipitously. Clearly these factors have the potential to have enormous impact on our economy. Indeed they appear to be happening in 2011. The near end of economic growth, the fact that in the United States some forty-six of fifty state governments are broke, that most European governments and the US federal government are having to restrict their expenditures while dealing with enormous deficits, that many pension plans have collapsed, that most universities are having severe financial problems—and so on—are all reflecting the cessation of growth in global petroleum production (Murphy and Hall 2011). If these are all as closely linked as David Murphy and Charles Hall believe, then we will have some very serious rethinking of our aspirations, goals, objectives, and, perhaps most importantly, how to think about sharing whatever economic pie is left.

Alternatives to Petroleum

The world is not about to run out of hydrocarbons, and perhaps it is not going to run out of oil any time soon. What will be difficult to obtain is cheap, high quality petroleum, because what is left is an enormous amount of low-grade hydrocarbons, which are likely to be much more expensive financially, energetically, politically, and especially environmentally. As conventional oil becomes less important, society has a great opportunity to make investments in a different source of energy, one freeing us for the first time from our dependence on hydrocarbons.

At the moment we do not have an energy shortage, as coal, gas, and perhaps nuclear sources remain reasonably abundant. But we do have a severe shortage of liquid fuel, which basically means oil. There are alternatives to conventional oil, which in the twenty-first century means unconventional oil, natural gas liquids, and ethanol. These can be mixed with oil pre- or postrefinery and contribute to our liquid fuel base. As of 2011 they supply about 10 percent of the world's liquid fuels. There are large costs, however. While tar sands (basically "undercooked" conventional oil) are abundant in Canada, Venezuela, and elsewhere—and are currently being exploited in Canada—their EROI is low, perhaps 3:1. The available quantity of this "undercooked" fuel will probably be restricted for some time, not by total supply but by the water, natural gas, and land required for processing. As of 2011, ethanol and natural gas liquids (NGL) together supply about 10 percent of world liquid fuels. These fuels, however, contain only about 70 (ethanol) to 73 (NGL) percent of the energy per volume that one obtains from oil, so that including them (as barrels) with the number of oil barrels in official statistics gives a somewhat inflated view of their importance. Another problem for ethanol is that, at least in the United States (the largest grower), the energy required to grow, harvest, and distill ethanol from corn is very nearly the quantity of energy in the fuel, so ethanol delivers little or no net energy (Murphy and Hall 2011). Hence these alternatives to oil have substantially lower quality than oil. There are vast reserves of what is called oil shale, which is essentially "overcooked oil," Numerous efforts have been made to develop them, but all have ended in failure for economic or other reasons. In 1970 they were supposedly going to be economically feasible to develop when oil hit $4 a barrel. In 2011, with oil at about $80–90 a barrel they are still not. Why? As the price of oil increases so does the provision of all goods and services in society, including those required to generate shale oil or any other alternative. This is a general reason that it has proved very difficult to come up with real alternatives to oil.

Besides oil and gas the principal supply of energy for the world is coal, which contributes roughly 25 percent. Hydroelectric power and wood are renewable energies generated from current solar input; each provides about 10 percent of the energy thc world uses. "New renewables" including windmills and photovoltaics, provide much less than one percent, and are not growing as rapidly in magnitude as petroleum (although they are as a percentage of their own contribution). The annual increase in oil and gas use is thus much greater than the new quantities coming from the new renewables, at least as of 2011, and they are not displacing fossil fuels but just adding to the mix. All of these proportions have not changed very much since the 1970s in the United States, Europe, or the world.

Outlook for the Future

In the early twenty-first century liquid and gaseous petroleum remain the foundation of our economies and our lives. Petroleum is especially important due to the magnitude of its current use; because it has important and unique qualitative attributes leading to high economic utility, including very high energy density and transportability; and because its future supply is worrisome. The issue is not the point at which oil actually runs out but rather the relation between supply and potential demand. Barring a massive worldwide recession, demand will continue to increase as human populations, petroleum-based agriculture, and economies (especially Asian) continue to grow. Petroleum supplies had been growing most years since 1900 at 2 or 3 percent per year, a trend that stopped in 2005 and that most investigators think cannot possibly continue. Peak oil, that is the time at which an oil field, a nation, or the entire world reaches its maximum oil production and then declines, is not an abstract issue debated by theoretical scientists or worried citizens but an actuality that occurred in the United States in 1970 and in some sixty (of ninety-five) other oil-producing nations since.

Several prominent geologists have suggested that it may have occurred already for the world.

There are several possible ways that humans can adjust to this new energy reality. Ideally we will be able to generate a new suite of energy technologies, including possibly a much less energy-intensive economy, that will enable civilization as we know it (or would like to), evolve to continue. Likewise there is certainly sufficient solar energy available to run human civilization. The trick is capturing it with a sufficient energy profit (i.e., EROI). How well we weather this coming storm will depend in large part on how we manage our investments now. From the perspective of energy there are three general types of investment that we make in society. The first is an investment in obtaining energy itself, the second is an investment for maintaining and replacing existing infrastructure, and the third is discretionary expansion. In other words, before we can think about expanding the economy we must first make an investment into getting the energy necessary to operate the existing economy, and also into maintaining the infrastructure that we have—at least unless we wish to accept the entropy-driven degradation of what we already have. Investors must accept the fact that the required investments into the second and especially the first category are likely to increasingly limit what is available for the third. The dollar and energy investments necessary for getting the energy that allows the rest of the economy to operate and grow have been very small historically, but this is likely to change dramatically, whether we continue our reliance on ever-scarcer petroleum or whether we attempt to develop some alternative. Technological improvements, if indeed they are possible, are extremely unlikely to bring back the low investments in energy that we have grown accustomed to.

The main problem that we face is a consequence of the "best first" principle. This is, quite simply, the characteristic of humans to use the highest quality resources first, be they timber, fish, soil, copper ore or, of relevance here, fossil fuels. This is because economic incentives are to exploit the highest quality, least cost (both in terms of energy and dollars) resources first, as was noted two hundred years ago by the British economist David Ricardo (1722–1823).

We have been exploiting fossil fuels for a long time. The peak in finding oil was in the 1930s for the United States and in the 1960s for the world, and both have declined enormously since then. An even greater decline has taken place in the efficiency with which we find oil, meaning the amount of energy we find relative to the energy we invest in seeking and exploiting it.

The Native Americans of the United States' plains depended completely on bison for their food, their lodging, their clothes, their implements, and their culture. They understood and acknowledged this, and devoted a considerable amount of their time to appreciating and even worshiping the bison through dance, stories, and thanks given to the Great Spirit and the bison themselves. Contemporary Americans (and Europeans and Asians and Pacific Islanders and just about everyone else) are similarly dependent upon petroleum, but we have devised a culture in which the credit for our affluence is given to human invention, to innovation, to entrepreneurship, to market economics and so on. While these factors certainly play a role, underneath and enabling it all has been our increasing use of petroleum, a factor hardly understood or appreciated. Even now political leaders tend to get the blame for the recent economic malaise that is almost certainly driven, in large part, by the end of the growth in our access to petroleum. Whatever the future of petroleum, and it is likely to be increasingly constricted, a failure to understand its critical role in our global society and the world economy is a recipe for disaster.

Charles A. S. HALL
State University of New York, Syracuse

See also in the *Berkshire Encyclopedia of Sustainability* Aquifers; Bioenergy and Biofuels; Coal; Geothermal Energy; Hydrogen Fuel; Industrial Ecology; Natural Gas; Natural Resource Economics; Solar Energy; Thorium; Uranium; Water Energy; Wind Energy

Author's note: This article is modified from Charles A. S. Hall and Kent Klitgaard (2011), *Energy and the Wealth of Nations: Understanding the Biophysical Economy.* New York: Springer. There is more supportive material in that source. I thank the Santa Barbara Family Foundation for support of much of the research behind this paper, as well as the UK Department for International Development for supporting the work of the State University of New York on issues related to the global energy rate of return and its implications for developing countries.

Further Reading

Bartlett, Albert A. (2000). An analysis of US and world oil production patterns using Hubbert-style curves. *Mathematical Geology 32, 1–17.*

Brandt, Adam R. (2007). Testing Hubbert. *Energy Policy 35*(May), 3074–3088.

Campbell, Colin J., & Laherrère, Jean H. (1998). The end of cheap oil. *Scientific American* 278, 78–83.

Cleveland, Cutler J.; Costanza, Robert; Hall, Charles A. S.; & Kaufmann, Robert. (1984). Energy and the United States economy: a biophysical perspective. *Science* 225, 890–897.

Energy Information Administration (EIA). (2000). Long term world oil supply. Retrieved September 25, 2011, from http://www.eia.gov/pub/oil_gas/petroleum/presentations/2000/long_term_supply/index.htm

Gagnon, Nate; Hall, Charles A. S.; & Brinker, Lysle. (2009). A preliminary investigation of energy return on energy investment for global oil and gas production. *Energies 2*(3), 490–503.

Guilford, M. C.; Hall, Charles A. S.; O' Connor, Peter; & Cleveland, Cutler J. (in press). A new long term assessment of EROI for US oil and gas discovery and production. *Sustainability.*

Hall, Charles A. S. (Ed). (in press). Special issue: New studies of EROI (Energy return on investment). *Sustainability*.

Hall, Charles A. S.; Cleveland, Cutler J.; & Kaufmann, Robert. (1986). *Energy and resource quality: The ecology of the economic process*. Wiley Interscience.

Hall, Charles A. S., & Cleveland, Cutler J. (1981). Petroleum drilling and production in the United States: Yield per effort and net energy analysis. *Science* 211, 576–579.

Hall, Charles A. S.; Lindenberger, Dietmar; Kummel, Reiner; Kroeger, Timm; & Eichhorn, Wolfgang. (2001). The need to reintegrate the natural sciences with economics. *BioScience* 51, 663–673.

Hall, Charles A. S., & Klitgaard, Kent. (2011). *Energy and the wealth of Nations: Understanding the biophysical economy*. New York: Springer.

Hall, Charles A. S.; Tharakan, Pradeep; Hallock, John; Cleveland, Cutler J.; & Jefferson. Michael. (2003). Hydrocarbons and the evolution of human culture. *Nature* 426, 318–322.

Hall, Charles A. S.; Powers, Robert; & Schoenberg, William. (2008). Peak oil, EROI, investments and the economy in an uncertain future. In David Pimentel (Ed.), *Biofuels, solar and wind as renewable energy systems: Benefits and risks*. Dordrecht, The Netherlands: Springer.

Hubbert, M. King. (1962). *Energy resources (A report to the Committee on Natural Resources)*. Washington, DC: National Academy of Sciences.

Kaufmann, Robert K., & Shiers, L. D. (2008). Alternatives to conventional crude oil: When, how quickly, and market driven? *Ecological Economics* 67, 405–411.

Murphy, David; Hall, Charles A. S.; & Powers, Robert. (2010). New perspectives on energy return on (energy) invested (EROI) of corn ethanol. *Environment, Development and Sustainability* 13,179–202.

Murphy, David, & Hall, Charles A. S. (2011). *Energy and the wealth of nations: Understanding the biophysical economy*. New York: Springer.

Nashawi, Ibrahim Sami; Malallah, Adel; & Al-Bisharah, Mohammed. (2010). Forecasting world crude oil production using multicyclic Hubbert model. *Energy Fuels* 24, 1788–1800.

Tissot, B. P., & Welt, D. H. (1978). *Petroleum Formation and Occurrence*. New York: Springer-Verlag.

United States Geological Survey (USGS). (2000). US Geological Survey world petroleum assessment 2000. Retrieved September 25, 2011, from http://pubs.usgs.gov/dds/dds-060/

Polluter Pays Principle

The polluter pays principle has become one of the most prominent standards on which worldwide environmental policy is based. The initial concept was first addressed in the 1970s; today its scope is much broader, encompassing not only pollution prevention and control but also liability for cleanup costs. In more recent years it has been extended to product impacts during the whole lifecycle.

The polluter pays principle (PPP) was first mentioned in the recommendation of the Organisation for Economic Co-operation and Development (OECD) of 26 May 1972 and reaffirmed in the recommendation of 14 November 1974. As a main function of the principle, these recommendations specify the allocation "of costs of pollution prevention and control measures to encourage rational use of scarce environmental resources and to avoid distortions in international trade and investment." The polluter should bear the expense of carrying out the measures "decided by public authorities to ensure that the environment is in an acceptable state" (OECD 1972).

In the 1972 Declaration of the United Nations Conference on the Human Environment in Stockholm, the principles did not feature, but in 1992 in Rio de Janeiro, PPP was laid down as Principle 16 of the UN Declaration on Environment and Development. This stated that national authorities should endeavor to promote the internalization of environmental costs and the use of economic instruments, taking into account the approach that the polluter should, in principle, bear the cost of pollution, with due regard to the public interest and without distorting international trade and investment.

The European Community took up the OECD recommendation in its first Environmental Action Program (1973–1976) and then in a recommendation of 3 March 1975 regarding cost allocation and action by public authorities on environmental matters. Since 1987 the principle has also been enshrined in the Treaty of the European Communities and in numerous national legislations worldwide.

Functions and Substance of PPP

Since its first appearance in 1972, the PPP is understood in a much broader sense today, not only covering pollution prevention and control measures but also covering liability, for example, costs for the cleanup of damage to the environment (OECD 1989). Also, the field of application of PPP has been extended in recent years from pollution control at the source toward control of product impacts during the whole lifecycle (known as extended producer responsibility, or EPR).

The preventive function of the PPP is based on the assumption that the polluter will reduce pollution as soon as the costs that he or she has to bear are higher than the benefits anticipated from continuing pollution. As the costs for precautionary measures also have to be paid by the potential polluter, he or she has an incentive to reduce risks and invest in appropriate risk management measures. Finally, the PPP has a curative function, which means that the polluter has to bear the cleanup costs for damage already occurred.

Since its overall objective is to make the polluter pay, the principle leads to the question of who the polluter is; this cannot be defined without knowing what pollution is.

What Is Pollution?

There are two different concepts for defining pollution: one is to establish administrative thresholds in order to define necessary preventive measures and environmental damages. If these thresholds are exceeded, there is pollution. In this view, pollution is congruent to unlawful acts. The second concept defines pollution independently from established thresholds and focuses only on the damage (or the risk of damage), that is, the environmental impact of the emission or harmful activity. As civil liability is not connected to the breach of administrative standards in most European legislations, the second concept is also more consistent with traditional legal concepts. The weakness of this approach is that PPP cannot provide an answer to the question of whether an impact is harmful or has to be considered as damage; it remains a challenge to natural and environmental sciences to define relevant criteria that then also could be implemented by legal standards. Insofar, both concepts do not necessarily contradict each other.

The polluter pays principle does not only apply if there is a "real" pollution in terms of harm or damage to private property and/or the environment. Most legal orders go beyond this interpretation: in light of the precautionary principle, environmental legislation may also provide for measures that are taken to minimize risks—even in cases where there is a lack of scientific knowledge and scientific cause–effect relationships cannot fully be established. One example is Article 3(3) of the United Nations Framework Convention on Climate Change (UNFCCC): "Where there are threats of serious or irreversible damage, lack of full scientific certainty should not be used as a reason for postponing such measures, taking into account that policies and measures to deal with climate change should be cost-effective so as to ensure global benefits at the lowest possible cost." In these cases, the responsible person (the plant operator, the producer of a product) has to bear the costs of precautionary measures according to the PPP, even though pollution has not yet occurred.

Who Is the Polluter?

The term *polluter* refers to a polluting, harmful activity. But the above-mentioned extension of the polluter pays principle has had the inevitable consequence that legislation today often defines the polluter in a more extensive way. Not only those polluters who, in a strict sense, actually "pollute" have to be considered as such, but also those who are only causing risks for the environment and where pollution has not yet occurred.

As far as polluted sites are concerned (for instance under the US Comprehensive Environmental Response, Compensation, and Liability Act, or CERCLA), the owners, operators of disposal facilities, generators of any hazardous wastes discovered at the polluted site, and transporters of the waste can all be considered polluters in this wider sense of the term. Under the German Soil Protection Act even former owners of the polluted site may be held liable under certain circumstances.

In the *Erika* oil spill case, the European Court of Justice held in 2008 that based on Article 15 of the European Union (EU) Waste Framework Directive, the producer of hydrocarbons that became waste due to the accident at sea could be held liable for the cleanup costs. (In 1999, the oil tanker *Erica* broke in two and polluted about 400 kilometers of the French coastline in Brittany.) In accordance with the polluter pays principle, however, such a producer is not liable unless he has contributed through his conduct to the risk of pollution stemming from the shipwreck.

The question of whether the "user" could also be regarded as a "polluter" is relevant, particularly in the field of product control law. Users often pay indirectly when pollution control costs are internalized in the prices of the product.

How Much Is Paid?

The polluter has to pay the costs for preventive and precautionary measures, administrative procedures and, in case of damage occurred, the costs for reinstatement. Although, from an economic point of view, the aim would be to achieve full internalization of external costs, in legal terms the responsibility of the polluter is limited by another general principle of law: the principle of proportionality. The extent to which preventive measures can be required by an operator depends on the risk at issue and the costs of the concrete measure.

A further restriction of the principle is that the polluter has to pay only the costs for his or her own pollution and not costs caused by other polluters, as the European Court of Justice held in the 1999 case (C-293/97) concerning the obligations of farmers under the European Union (EU) Council Implementation of Nitrates Directive (91/676/EEC) to reduce the concentration of nitrates in waters below a determined threshold.

There are some exceptions to this rule, for example, when the law provides for joint and several liability. This means that in cases of several polluters, the injured party can claim for total compensation against one of the polluters of his or her choice (as under CERCLA). Under several regimes of strict liability, the maximum amount for which the polluter is liable in the case of damage is limited, whereas under the general law of tort there is no such limit.

PPP and Other Environmental Principles

As we have seen, the PPP is closely linked to other environmental principles, such as the prevention and the precautionary principles. As a complementary principle to the latter, the PPP has evolved into a comprehensive principle of polluter responsibility.

One particular form of PPP is the extended producer responsibility (EPR). EPR is defined by the OECD as the extension of responsibility to a postconsumer stage of a product's lifecycle both physically and economically. The aim is to provide incentives for producers to improve product design in terms of sustainability. This principle is applied, for instance, in the European Waste Electronic and Electrical Equipment Directive (WEEE), in the End of Life Vehicles Directive, and in the Battery Directive. Also the new EU Waste Framework Directive (2008/98/EC) provides for EPR. This states that EU member states may take appropriate measures to encourage the design of products in order to reduce their environmental impacts and the generation of waste in the course of the production and subsequent use of products, and to ensure that the recovery and disposal of products that have become waste take place without endangering human health and the environment. Measures under EPR may encourage the development, production, and marketing of products that are suitable for multiple use, that are technically durable, and that are suitable for proper and safe recovery and environmentally compatible disposal. EPR is limited by technical and economic feasibility. In the United States, one type of EPR is also known as product stewardship, which is an approach that is based more on voluntarism and shared responsibility of all stakeholders involved (but the producer still has a potentially larger responsibility).

The concept of individual producer responsibility (IPR) is laid down in the WEEE Directive. The main incentive established by the directive is the allocation of costs. Under the directive every producer of electronic or electrical equipment shall be responsible for financing the operations relating to the waste from their own products. The take-back systems set up in the EU member states are organized more collectively, however; the producer is responsible for an undifferentiated mixture of devices and not only for his own products (called collective producer responsibility). Thus, there are some shortcomings in the implementation of IPR in practice.

One understanding of the PPP interprets the principle only as a policy principle with no legal impact. This interpretation is based on the fact that the principle as such is vague and therefore not legally enforceable without further concretization. By contrast, another understanding posits that vague legal concepts are not unusual in law. Since 1972 the principle has been introduced in numerous national and international legal texts and therefore is generally considered today as a legal principle, although its application in practice depends on the implementation of further instruments.

PPP in International Law

PPP is recognized in a number of international conventions (most of which have a regional character) like the Helsinki Convention on the Protection of the Baltic Sea or the Barcelona Convention for the Protection of the Mediterranean Sea against Pollution. There is not yet a unanimous opinion as to whether PPP should be considered as a general principle of law or as a rule of customary law as provided for in Article 38 of the Statute of the International Court of Justice. The fact that most nations have introduced PPP into their national legal orders indicates there is growing international acceptance for it. This along with an increasing number of international conventions that refer to it are both strong arguments in favor of the reconnaissance of PPP as a general principle of law.

General acceptance of PPP can also be observed under World Trade Organization (WTO) law and that of its predecessor, the General Agreement of Tariffs and Trade (GATT).

PPP in European Law

Since 1987 the European Community (EC) Treaty has provided that Europe's policy on the environment shall be based, among other things, on the principle that the polluter should pay (formerly Article 174, paragraph 2 of the EC Treaty; currently Article 191, paragraph 2 of the Treaty on the Functioning of the European Union, or TFEU). Without further definition of the principle, the TEEU obliges the nations to implement the PPP in their environmental policy. Although we may conclude from this that PPP has a legally binding effect, the significance of this effect is controversial. In a case dealing with the prohibition of the use of hydrochlorofluorocarbons (HCFCs), the European Court of Justice held that the objectives, principles, and criteria set out in the relevant article of the EC Treaty on the Environment (1998) have to be respected by the legislature in implementing that policy. But the court limited its own review competence in saying that review by the court must necessarily be limited to the question of whether, in adopting particular rules, the legislature committed a manifest error of appraisal regarding the conditions for the application of the relevant article of the treaty providing for PPP.

In fact, numerous directives and regulations—so-called secondary legislation—have put into force a widespread corpus of environmental law that is largely based on the PPP.

PPP in National Law

Environmental legislation across the world is increasingly recognizing PPP. In some countries (e.g., Australia), specialized environmental courts have been set up to control the enforcement of environmental legislation and PPP. In Canada, PPP is enshrined in the preamble of the Environmental Protection Act 1999; in France the Environmental Code of 2000 defines PPP in Article L.110-1 as a principle under which the costs for preventive measures as well as reduction and remedying measures have to be borne by the polluter.

In developing and emerging countries new environmental legislation is frequently inspired and set up—often with the support of multilateral or bilateral international cooperation—by modern concepts containing environmental principles such as the PPP.

Implementation and Fields of Application

To be effective, the PPP has to be implemented by concrete instruments in international and national legislation. There are a number of legal and economic instruments available to this end. These instruments can also be combined and do not exclude each other, given that full cost internalization is usually not reached by a single instrument alone.

Environmental binding standards, emission limit values, or the so-called "best available technique" approach, set up and defined by government, are still predominant tools in many countries and environmental sectors, such as air and water pollution control (e.g., in the US Clean Air Act of 1970 and the US Clean Water Act of 1977) or cleanup of contaminated sites. According to the PPP, the costs for meeting these standards (for instance the investment of "depollution" techniques) have to be paid by the plant operators.

In the last two decades, economic instruments have gained more relevance. Tradable permits try to provide operators with an incentive to invest in pollution control measures in the most efficient way. Title IV of the US Clean Air Act Amendments pursued the objective of reducing national sulfur dioxide emissions from electricity utilities by 50 percent between 1995 and 2000 during Phase I. The Kyoto Protocol has established an international trading scheme for carbon dioxide emissions, which has also been implemented into European law. It is questionable, however, whether emission trading programs have an advantage over classic command-and-control law. This particularly holds true if allowances for existing plants are distributed free of charge ("grandfathering"), as was the case in the first trading period under the European carbon dioxide emissions trading program.

Several European countries, such as Belgium and Germany, have introduced ecotax (ecological taxation) schemes in their respective tax legislation in order to promote a more environmentally friendly use of raw materials or energy. Cost recovery charges have been set up in Australia for the purpose of financing administrative activities in the fishery sector. Furthermore, new liability rules have been set up in many countries and in the European Union under the 2004 EU Environmental Liability Directive. Certain professional activities fall under a strict liability regime if environmental damage to soil, water, or natural habitats and species is caused. The responsible person then has to inform the authorities, prevent further harm, and reinstate the damaged environment.

Another instrument is the concept of property rights that seeks to give ownership to individuals of unowned natural goods.

PPP's Contribution to Sustainability

Today the polluter pays principle has become one of the most prominent standards on which environmental policy is based. It influences not only environmental policy but also environmental law on international, European, and national levels. In particular, environmental legislation in the European Union is increasingly based on PPP. The PPP has contributed considerably to the success achieved in improving air and water quality in Europe in the last thirty years.

The worldwide acceptance of PPP will probably also lead to its acceptance as a general principle of international law in the near future. Important accidents like the 2010 oil spill in the Gulf of Mexico show the constant and ever-increasing relevance of the PPP. Thus, together with the principles of prevention and precaution, the PPP will remain a core basis for sustainable development.

Gerhard ROLLER
University of Applied Sciences

See also in the *Berkshire Encyclopedia of Sustainability* Bhopal Disaster; Civil Liability Convention for Oil Pollution Damage; Clean Air Act; Clean Water Act; Convention for the Prevention of Pollution From Ships; Convention on Long-Range Transboundary Air Pollution; Convention on Persistent Organic Pollutants; Environmental Law (several articles: Australia and New Zealand; Europe; United States and Canada); Precautionary Principle; Transboundary Water Law; Waste Shipment Law

Further Readings

Beder, Sharon. (2006). *Environmental principles and policies: An interdisciplinary introduction*. London: Earthscan.

de Sadeeler, Nicolas. (2002). *Environmental principles*. Oxford, UK: Oxford University Press.

Epiney, Astrid. (2006). Environmental principles. In Richard Macrory (Ed.), *Reflections on 30 years of EU environmental law*. Groningen, The Netherlands: Europa Law Publishing.

Krämer, Ludwig. (1997). Polluter-pays-principle in community law: The interpretation of article 103r of the EEC Treaty. In Ludwig Krämer (Ed.), *Focus on European Law* (pp. 244). London: Graham & Trotman.

Mann, Ian. (2009). A comparative study of the polluter pays principle and its international normative effect on pollutive processes. Retrieved July 13, 2010, from http://www.consulegis.com/fileadmin/downloads/thomas_marx_08/Ian_Mann_paper.pdf

Renckens, Stefan. (2008). Yes, we will! Voluntarism in US e-waste governance. *RECIEL, Review of European Community & International Law, 17*(3), 286–299.

Roller, Gerhard, & Führ, Martin. (2008). Individual Producer Responsibility: A remaining challenge under the WEEE Directive. *RECIEL, Review of European Community & International Law, 17*(3), 279–285.

Treaties / Resolutions / Court Cases

Council of the European Communities Directive of 12 December 1991 concerning the protection of waters against pollution caused by nitrates from agricultural sources (91/676/EEC). Retrieved August 28, 2010, from http://ec.europa.eu/environment/water/water-nitrates/directiv.html

Council of the European Communities Waste Framework Directive (2008/98/EC) of 19 November 2008.

Organisation of Economic Cooperation and Development (OECD). (1972, May 26). Recommendation of the council on guiding principles concerning international economic aspects of environmental policies. Council Document no. C(72)128.

Organisation for Economic Co-operation and Development (OECD). (1974, November 14). Recommendation of the council on the implementation of the polluter-pays principle. Document no. C(74)223.

Organisation for Economic Co-operation and Development (OECD). (1989, July 7). Recommendation of the council concerning the application of the polluter-pays principle to accidental pollution. Document no. C(89)88/FINAL.

International Court of Justice. (1945). Statute of the International Court of Justice. Retrieved November 15, 2010, from http://www.icj-cij.org/documents/index.php?p1=4&p2=2&p3=0

United Nations Framework Convention on Climate Change (adopted on 9 May 1992 and entered into force 21 March 1994).

Shale Gas Extraction

Technologies for extracting gas from shale by using high-volume hydraulic fracturing (commonly known as "fracking"), a process that breaks up the rock and releases the gas, have developed since about the year 2000. Although many see shale gas as a viable alternative to other fossil fuels, especially natural gas from conventional sources, the environmental costs are high. Particular concerns include water and air pollution as well as emissions of greenhouse gases.

Natural gas makes up some 20 percent of all energy use globally (IEA 2011) and 24 percent in the United States (US EIA 2011). Most natural gas is obtained by drilling a well into a pocket of gas trapped beneath an impermeable layer within the Earth. The gas simply flows through the well to escape and reach the surface. People have gradually been depleting such sources of "conventional" natural gas, however, and are increasingly turning to "unconventional" sources: gas that is tightly held in rocks with very low permeability such as shale, some sandstones, and coal seams. High-volume hydraulic fracturing (fracking) is a controversial method that is used to extract natural gas from unconventional sources.

The fracking process involves forcing water or other fracturing fluid into a gas or oil wellbore to create fractures and small fissures in the underlying rock, which will release the gas and thus increase the production of the wells. The technique has been used since the 1940s in conventional gas and oil wells, but until recently, only with modest quantities of water—a few hundred thousand liters of water per fracking event per well at most. (Exactly how often industry may re-frack wells, or if they do so at all at significant levels, remains unknown, as the technology is simply too new.) The industry started experimenting with high-volume hydraulic fracturing of shales in Texas in the mid-1990s, but using relatively low volumes, and relatively few wells. By 2003 or so, they were using larger volumes, and starting to use the approach for significant production of shale gas, but still only in Texas. They started later in that decade to move into other states, and to increase the volumes used. Mostly, this practice is still quite new. For instance, in the Marcellus shale in Pennsylvania, significant shale gas extraction began only in 2009.

To get unconventional gas from shales, producers began to use much larger volumes of water as well as many chemical additives, combined with high-precision horizontal drilling of wells. With high-precision drilling, workers can bore down into the Earth to depths of 3 kilometers or more, then curve the well and drill sideways for another 2 kilometers or more, closely following within a vein of a particular gas-rich shale. (See figure 1 on page 138.) The shale rock is then fractured by forcing large volumes of water—20 million liters of water per well on average—and additives through the well at high pressure. In addition to the chemical additives discussed in the section "What Goes Down / What Comes Back Up," many of which have not been identified to the public, sand or other fine particles are included in the mixture to help keep the new fissures open to aid in the flow of gas from the rock and up the well.

Shale Gas Is New

The development of unconventional gas by fracking is fairly new, having started slowly with gas in tight sandstones in the 1980s, then in coalbeds in the early 1990s, and in shale in the late 1990s. In the United States, production of gas from tight sandstones and coalbeds has already peaked, and only shale gas production is expected

Figure 1. Drilling for Gas from Conventional and Unconventional Sources

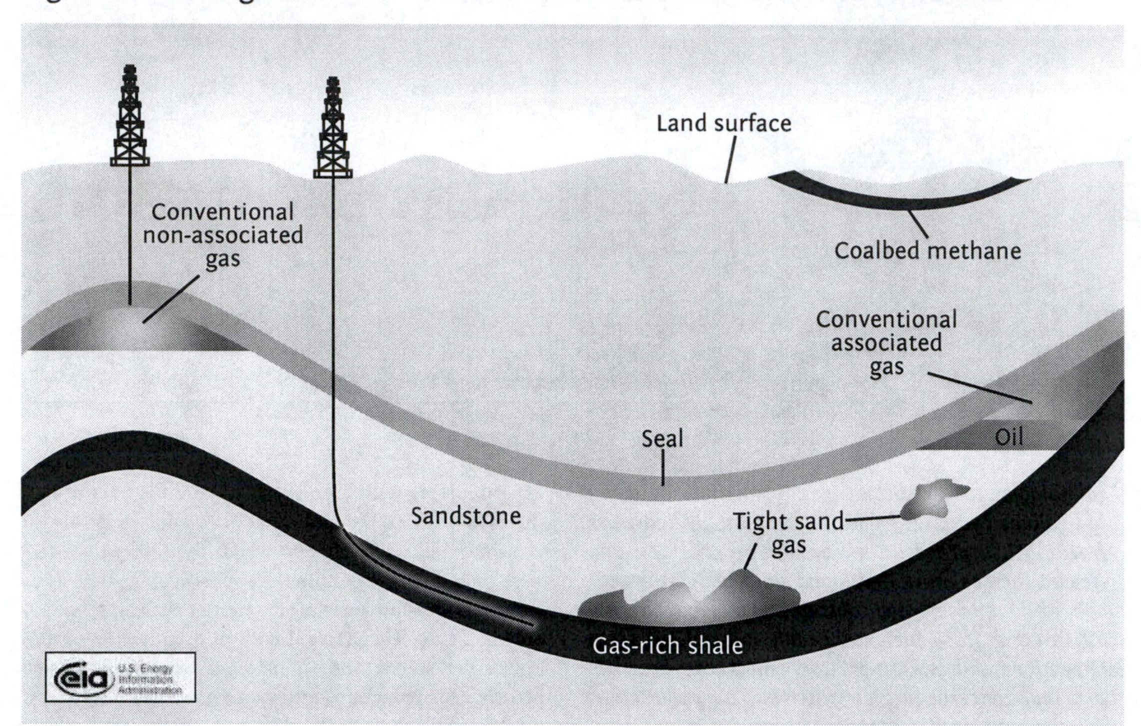

Source: Energy Information Administration of the US Department of Energy (US EIA 2010).

For conventional natural gas, a well is drilled into a formation where the gas is trapped, and the gas readily flows up the well to the surface. For the tightly held natural gas in shales, wells are drilled directionally within the vein of rock. The rock is then hydraulically fractured to create fissures, releasing the gas and allowing it to flow.

to increase over the coming decades (US EIA 2011). Shale gas development began in Texas, and Texas still dominates the production of shale gas globally, although some commercial production came online in Arkansas, Louisiana, and Pennsylvania between 2007 and 2009. In 2007, shale gas contributed only 1 percent of the supply of natural gas to the United States. By 2009, this had grown to 14 percent, and the US Department of Energy has projected that shale gas will supply 45 percent of the country's natural gas by 2035 (US EIA 2011), although many researchers believe this is too optimistic (Howarth and Ingraffea 2011). Outside of the United States, exploratory shale gas wells have been drilled in Quebec and British Columbia, Canada, and in a few European countries, but no shale gas wells have gone into commercial production in any of these places. Many parts of the world contain possible shale gas resources, and some researchers predict a massive global explosion in shale gas development (Engelder 2011). Geologists do not know precisely how large the world's reserves of shale gas are, however, and a report by the US Geological Survey in August 2011 (Coleman, J. L. et al. 2011) cast doubt on the more optimistic estimates (Howarth and Ingraffea 2011).

Scientific study of shale gas extraction's environmental effects is in the early stages. In fact, the first studies of the environmental consequences of shale gas were published in 2011 (Howarth and Ingraffea 2011). In 2005, the US Congress exempted fracking from most federal environmental oversight and regulation, which has made it difficult to obtain information on fracking's effects from industry sources. Nonetheless, a growing body of evidence indicates reason for concern.

What Goes Down/What Comes Back Up

In addition to the huge volume of water used, roughly 200,000 liters of chemicals are added to a well during the fracking process. These include acids to assist with opening up fissures, biocides to prevent microorganisms from growing and clogging up the fissures, scale inhibitors to reduce corrosion of the pipes, and surfactants to reduce the friction of the high volume of water at high pressure traveling through the long pipe runs of the well. Many of these chemical additives are toxic, mutagenic (causing birth defects), or carcinogenic (causing cancer). The exact composition of the additives used has not been disclosed to the public because the 2005 congressional fracking exemption allows the industry to keep the chemical list secret (Howarth and Ingraffea 2011). Similarly, Canada has no requirements about disclosing the composition of fracking chemicals (De Souza 2011). Nonetheless, although disclosure is not the industry standard, more drilling companies are voluntarily reporting their chemical mixtures as the public becomes more vocal about obtaining the information.

When injected into the well, the frack water and chemicals extract additional materials from the rock formations, including toxic heavy metals, organic materials (including some such as benzene that are toxic and carcinogenic), and radioactive substances such as thorium, radium, and uranium. The fracking of a well generally takes less than a day. During the following two weeks or so, some of the water (approximately one-fifth of what was added) together with the additives and extracted materials, flows back to the surface. These mixtures are called *flow-back fluids*.

Treatment and Disposal of Flow-Back Fluids

When the flow-back fluids come to the surface, they are stored in open pits or in tanks until they are treated or disposed of. In Texas, the industry disposes of most of the flow-back fluids by injecting them into old, abandoned conventional oil and gas wells. Elsewhere in the United States, there are not enough abandoned wells to provide sufficient disposal capacity, and other approaches are necessary. In Pennsylvania, for example, most flow-back fluids have been trucked to municipal sewage treatment plants. Unfortunately, these facilities are not designed to handle the toxic materials, and much of the waste has simply flowed through the sewage plants and been discharged into rivers (Howarth and Ingraffea 2011; Urbina 2011a). In the summer of 2011, the State of Pennsylvania outlawed using sewage treatment plants for flow-back fluid disposal. The natural gas industry is attempting to develop effective and nonhazardous disposal methods, such as to recycle the fluids and reuse them in fracking. To date, only small percentages of flow-back fluids have actually been recycled (Urbina 2011b), and the future for waste disposal is highly uncertain.

Water Pollution

Improper disposal of flow-back fluids can lead to surface water pollution, and, in addition, the development of unconventional gas may contaminate groundwater. Freshwater aquifers are usually at shallow depths underground, within the top 100 meters or so, while the shale gas is at depths of a kilometer or more. Despite this distance, evidence indicates that in at least some cases, fracking fluids actually have entered surface aquifers (Urbina 2011d). One mechanism for the contamination may be leaks in the well pipes as they pass through the aquifer. Another possibility is that the high pressure used in fracking forces the fluids up through nearby older, abandoned wells, and these in turn leak into the groundwater aquifer. The contamination of groundwater by frack fluids has received little study or scrutiny, however, in part because information about it has been sealed and kept from the public when drilling companies settle lawsuits with landowners whose water has been contaminated (Urbina 2011d).

More common than contamination with fracking fluids is contamination with methane gas. A team of scientists from Duke University demonstrated high levels of methane contamination in many private drinking water wells within one kilometer of gas wells in Pennsylvania (Osborn et al. 2011). Water wells at greater distances from gas wells sometimes had methane contamination too, but at much lower concentrations. Natural gas is composed mostly of methane, and this study proved that the high levels of methane contamination came from the deep shale gas, and not from other sources of methane closer to the surface (such as bacteria in waterlogged soils). The Duke team did not find fracking fluid contamination in the water wells they sampled, and methane is not toxic. The methane did occur at levels that pose a major risk of explosion, however. Furthermore, the fact that methane could migrate from the deep shale formation into surface water wells suggests that other gases from the shale, such as benzene vapor, may also be migrating and contaminating the wells.

Air Pollution

Shale gas development is a major industrial enterprise that results in sometimes severe air pollution (Howarth and Ingraffea 2011). Large numbers of trucks haul water to the wells and flow-back fluids away. Massive diesel engines are used to drive the drills through kilometers of rock, and ten-thousand-horsepower diesel engines drive the pumps for the actual fracking. More engines run compressors to deliver the gas through pipelines. Toxic organic gases and vapors—compounds such as benzene and toluene—are routinely vented and leaked into the air. Cumulatively, these emissions can lead to high levels of ozone, which pose a risk to human health but also adversely affect the vegetation of natural ecosystems. Since shale gas drilling began in rural Colorado, ozone concentrations in the once pristine air have often approached or exceeded the regulatory standard set by the US Environmental Protection Agency (CDPHE 2010).

In Texas and Pennsylvania, state regulatory agencies routinely measure benzene concentrations in the air at levels that pose a significant risk of cancer from chronic exposure, and at times in Texas the concentrations exceed the acute public health standard (Howarth and Ingraffea 2011).

Greenhouse Gas Emissions

Shale gas has been widely promoted as a clean fuel, one with fewer greenhouse gas emissions than coal or oil, and therefore suitable as a bridge fuel that would let society continue to rely on fossil fuels while reducing global warming to some extent. Shale gas does indeed produce less carbon dioxide than coal or oil for an equivalent amount of energy, but this is only part of the emissions story. Methane is an incredibly powerful greenhouse gas—105 times more potent than carbon dioxide over a twenty-year period following emission (Shindell et al. 2009). Consequently, even small leakages of shale gas, which is mostly methane, have a huge influence on the greenhouse gas footprint of shale gas.

In April 2011, the first comprehensive analysis of emissions of all greenhouse gases from shale gas development, including methane as well as carbon dioxide (Howarth, Santoro, and Ingraffea 2011), evaluated the venting (purposeful emission) and leakage (accidental) of methane from the time of fracking and well completion through the processing of gas and delivery to the final consumer. The analysis found that shale gas development emits more methane than does conventional natural gas, due to a large venting of gas during the two-week flow-back period following fracking. Methane emissions dominate the greenhouse gas footprint of shale gas, giving this fuel a larger footprint than any other fossil fuel when considered over a twenty-year period. (See figure 2 for a comparison of greenhouse gas emissions for different types of fuel.) As time goes on, though, the influence of methane is diminished, as methane is removed from the atmosphere some ten times faster than is carbon dioxide. Nonetheless, the greenhouse gas footprint of shale gas is comparable to that of other fossil fuels over periods of up to one hundred years or more when used to generate heat (the primary use of natural gas) and over periods of up to fifty years when used to generate electricity (Howarth and Ingraffea 2011; Howarth, Santoro, and Ingraffea 2011, 2012; Hughes 2011). That is, shale gas extracted by fracking should not be used as a bridging fuel if society is to reduce global warming and avoid tipping points in the global climate system over the coming decades.

Figure 2 on page 141 illustrates the total greenhouse gas footprint of shale gas in comparison to other fossil fuels, considered at the integrated time scale for the 20 years following emissions. The footprints for each fossil fuel represented by the six columns in figure 2 are divided into three segments: (1) the direct emission of carbon dioxide from burning the fuel (as indicated by the bottom portion of each column); (2) indirect emissions of carbon dioxide necessary to develop and use the fuel, including for example trucking water to a fracking site and carrying coal in trains (as indicated by the small sliver in the middle of each column); and (3) methane emissions, converted to equivalents of carbon dioxide over a twenty-year integrated time period (as indicated by the top segment of each column). The figure provides both low and high estimates for methane emission rates from shale gas and from conventional gas as well as for surface and deep-mined coal. Note that while methane is emitted from coal mining and oil wells, the amount is small compared to the leakage from natural gas.

Figure 2. The Greenhouse Gas Footprint of Shale Gas

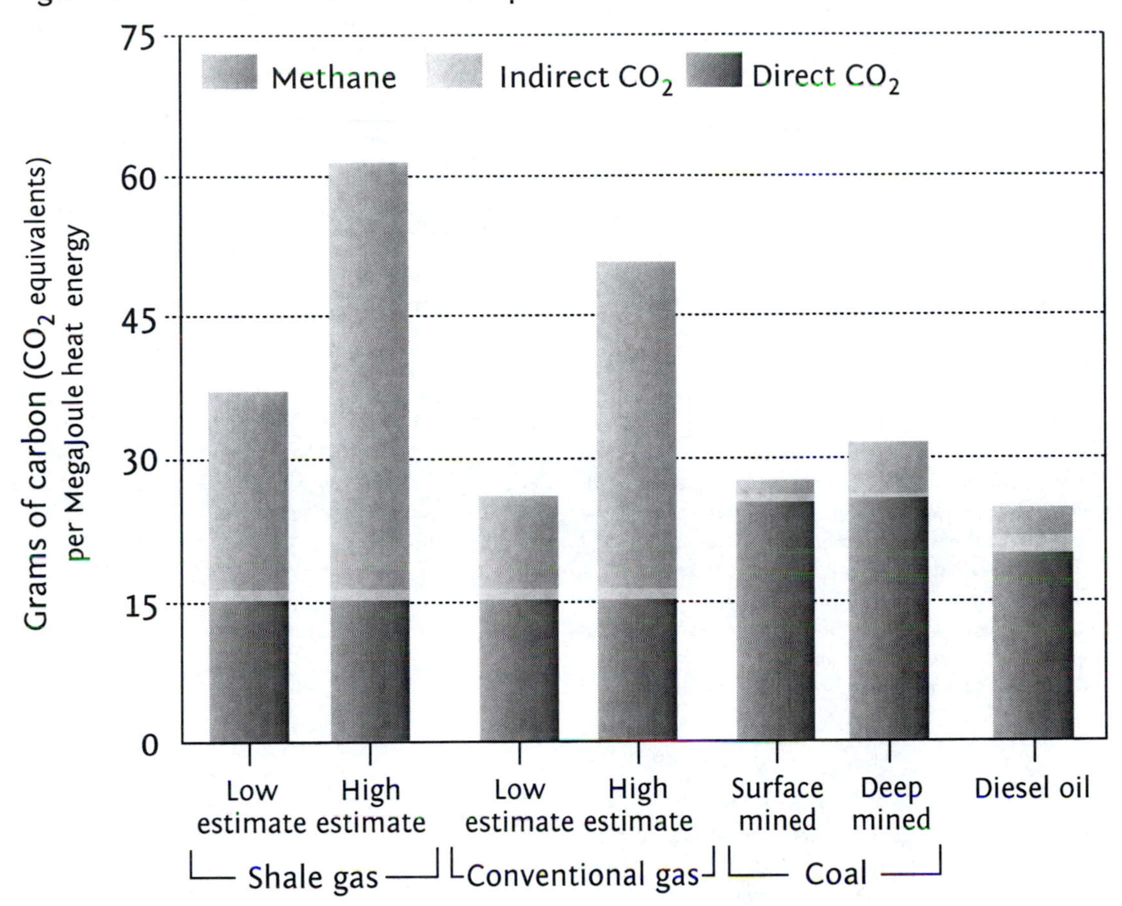

Source: Howarth and Ingraffea (2011).

The greenhouse effects of shale gas are predicted to exceed those of conventional gas, oil, and coal.

Shale Gas and the Environment

Shale gas development is relatively new, and the industry is still developing techniques and technologies. Research into reducing the environmental consequences of the process is ongoing. For instance, there is an urgent need to develop appropriate methods for treating and disposing of flow-back fluids. Whether this will happen, and whether new treatment technologies such as recycling the wastes can be done in a manner industry considers economical, remains to be determined. An important factor is that the cost of extracting the shale gas is high compared to the market price of natural gas (Howarth, Santoro, and Ingraffea 2011b; Urbina 2011c). From 2009 through 2011, natural gas prices hovered near four dollars per thousand cubic feet of gas, yet the break-even point for companies to turn a profit in developing shale gas is probably greater than six dollars per thousand cubic feet, but "if history is a guide, the cost of production of any new resource always drops over time" (RTEC n.d.), which means initial high costs may not be a deterrent to producers.

The technology for reducing methane emissions, thereby reducing the greenhouse gas footprint of shale gas, is well developed. The gas vented during the flow-back period can be captured and sold, rather than released to the atmosphere. But at current prices the value of the gas is small compared to the cost of the capture, giving industry at best a small return on investment. As a result, industry captures the gas less than 15 percent of the time when they complete wells (Howarth, Santoro, and Ingraffea 2012). Further, much of the methane emissions cannot be captured, because it is in the form of leaks from pipelines as the gas is pumped to consumers. In the United States, the average pipeline is more than fifty years old (CEQ 2004), and both long-distance transmission pipelines and distribution pipelines within cities can be quite leaky. The price of replacing these pipelines with modern technology is very high and of questionable value if the goal is to use shale gas as a bridge fuel for two or three decades before moving to truly green, renewable sources of energy.

Outlook

Shale gas is widely distributed across the planet. As society depletes conventional sources of natural gas and other fossil fuels, shale gas suggests to some the potential to continue to rely on fossil fuels over the coming decades. The environmental consequences are high, however, with widespread water and air pollution. Further, shale gas has a larger greenhouse gas footprint than any other fossil fuel, when evaluated over a period of fifty years or less following emission. As a result, reliance on this resource using existing extraction technology will tax the planet. Improvements in technology and better capture of wastes and emissions are necessary if shale gas is to be part of a sustainable future.

Robert W. HOWARTH
Cornell University

See also in the *Berkshire Encyclopedia of Sustainability* Groundwater Management; Hydrology; Pollution, Point Source; Waste Management; Water Resource Management, Integrated (IWRM)

Further Reading

Coleman, James L., et al. (2011). Assessment of undiscovered oil and gas resources of the Devonian Marcellus shale of the Appalachian Basin Province, 2011: US Geological Survey fact sheet 2011–3092. Retrieved December 8, 2011, from http://go.nature.com/8kejhm

Colorado Department of Public Health and Environment (CDPHE). (2010). Public health implications of ambient air exposures as measured in rural and urban oil & gas development areas: An analysis of 2008 air sampling data. Retrieved December 8, 2011, from http://go.nature.com/5tttna

The Council on Environmental Quality (CEQ), et al. (2004). Memorandum of understanding on coordination of environmental reviews for pipeline repair projects. Retrieved December 8, 2011, from http://www.fws.gov/habitatconservation/PSIA_MOU_FINAL_with_signatures_06_18_04.pdf

De Souza, Mike. (2011, October 25). Shale gas explorations a "game changer." *Vancouver Sun*. Retrieved on October 26, 2011, from http://www.vancouversun.com/business/Shale+exploration+game+changer/5601163/story.html

Engelder, Terry. (2011, September 15). Should fracking stop? No, it is too valuable. *Nature, 477*, 271, 274–275.

Howarth, Robert W., & Ingraffea, Anthony. (2011, September 15). Should fracking stop? Yes, it is too high risk. *Nature, 477*, 271–273.

Howarth, Robert W.; Santoro, Renee; & Ingraffea, Anthony. (2011a, April13). Methane and the greenhouse gas footprint of natural gas from shale formations. *Climatic Change*. doi: 10.1007/s10584-011-0061-5

Howarth, Robert W.; Santoro, Renee; & Ingraffea, Anthony. (2012, in press). Venting and leakage of methane from shale gas: Issues of gas use, time scale, economics, and regulation. *Climatic Change*.

Hughes, David. (2011, May 29). Will natural gas fuel America in the 21st century? Santa Rosa, CA: Post Carbon Institute. Retrieved October 26, 2011, from http://www.postcarbon.org/report/331901-will-natural-gas-fuel-america-in

International Energy Agency (IEA). (2011). Natural gas. Retrieved December 7, 2011, from http://www.iea.org/subjectqueries/keyresult.asp?keyword_id=4108

Osborn, Stephen G.; Vengosh, Avner; Warner, Nathaniel R.; & Jackson Robert B. (2011). Methane contamination of drinking water accompanying gas-well drilling and hydraulic fracturing. *Proceeding of the National Academy of Science, 108*, 8172–8176.

Pennsylvania Department of Environmental Protection. (2011). Northeastern Pennsylvania Marcellus shale short-term ambient air sampling report. Retrieved December 8, 2011, from http://go.nature.com/tjscnt

Research Triangle Energy Consortium (RTEC). (n. d.). Shale gas. Retrieved December 8, 2011, from http://rtec-rtp.org/shale-gas/

Shindell, Drew T.; Faluvegi, Greg; Koch, Dorothy M.; Schmidt, Gavin A.; Unger, Nadine; and Bauer, Susanne. (2009). Improved attribution of climate forcing to emissions. *Science, 326*, 716–718.

Texas Commission on Environmental Quality. (2010). Barnett shale formation area monitoring projects. Retrieved December 8, 2011, from http://go.nature.com/v7k4re

US Energy Information Administration (US EIA). (2010). Schematic geology of natural gas resources. Retrieved December 8, 2011, from http://www.eia.gov/oil_gas/natural_gas/special/ngresources/ngresources.html

US Energy Information Administration (US EIA). (2011). Annual energy outlook 2011: United States Energy Information Administration, Department of Energy, report #0383ER. Retrieved on October 26, 2011, from http://www.eia.gov/forecasts/aeo/

Urbina, Ian. (2011a, February 26). Regulation lax as gas wells' tainted water hits rivers. *New York Times*. Retrieved October 26, 2011, from http://www.nytimes.com/2011/02/27/us/27gas.html?hp

Urbina, Ian. (2011b, March 1). Wastewater recycling no cure-all in gas process. *New York Times*. Retrieved October 26, 2011, from http://www.nytimes.com/2011/03/02/us/02gas.html?hp

Urbina, Ian. (2011c, June 26). Behind veneer, doubt on future of natural gas. *New York Times*. Retrieved October 26, 2011, from http://www.nytimes.com/2011/06/27/us/27gas.html?ref=ianurbina

Urbina, Ian. (2011d, August 3). Tainted water well a concern, and there may be more. *New York Times*. Retrieved October 26, 2011, from http://www.nytimes.com/2011/08/04/us/04natgas.html?ref=ianurbina

Sugarcane

Sugarcane produces multiple harvests from a single planting and is an important source of sugar and bio-ethanol; the latter is a renewable alternative fuel that is produced in an energy-efficient way from the by-product of sugar manufacturing or squeezed sugar juice from stalks of sugarcane. Some of sugarcane's plant residues and by-products are used for animal feed, fertilizer, manufacturing, and energy production.

Sugarcane (*Saccharum* spp. hybrid) is a tall perennial grass that is cultivated widely in tropical and subtropical regions of the world because its stems contain juice with 10–15 percent sucrose at harvest. Sucrose is used in sugar and ethanol production. Sugarcane is a C_4 plant, meaning it creates 4-carbon molecules during photosynthesis. It also has a high photosynthetic ability, or rate; and it is one of the biggest biomass producers (and high-biomass energy sugarcanes are being developed). The plant belongs to the Poaceae family and the Panicoide subfamily. Wild sugarcane, *Saccharum officinarum*, is indigenous to New Guinea. Most commercial varieties of sugarcane are hybrids of *S. officinarum*.

Sugarcane is considered a sustainable crop because there is no decrease of crop yield even after continuous long-term cultivation. The main element of the final products—sugar and ethanol—are carbon fixed during photosynthesis, and most of the other elements can be returned to the field; plant residues and by-products of sugar and ethanol production act as fertilizers to maintain soil fertility.

The amount of area for sugarcane cultivation is increasing, and the total worldwide production (weight of fresh cane) was 1,743 million metric tons on 24.4 million hectares of harvested area in 2008 (FAO 2010). Brazil is the world's largest sugarcane producer, with one-third of the world's production, followed by India, China, Thailand, Pakistan, and Mexico.

Sugar beet (*Beta vulgaris* L.) is other source of sugar, and its big root contains 14–20 percent sucrose by weight. Commercial production of sugar beets began in the nineteenth century in Europe, and the biggest sugar beet producer is France, followed by Russia, the United States, and Germany. Total sugar production in the world is about 160 million metric tons. About 70 percent of the world's sugar production originates from sugarcane, and 30 percent from sugar beets. Major sugar exporters are Brazil, India, Thailand, and Australia.

An important crop in European colonies, sugarcane was planted in plantations. Crystallized sugar was invented in India around the fifth century and spread to China, Arabia, and Europe. During the seventeenth and eighteenth centuries, sugar from the Caribbean formed one side of the triangular trade, along with manufactures from Europe and slaves from Africa.

Cultivation

Sugarcane is a monoculture crop (one crop grown in a large area) and involves continuous cropping, which means it is grown on the same land each year. It is propagated by planting stem cuttings that have buds on them. Mechanical planting is not common except in some developed countries. The optimal temperature for sugarcane cultivation is between 20°C and 35°C. Because of its resistance to strong wind, it can be planted in areas that are affected by typhoons, hurricanes, and cyclones, such as the islands of Okinawa in Japan and the Caribbean. Lower temperatures, drought periods, or both can trigger sucrose accumulation in the plants. Brazil has the largest potential land area to increase sugarcane production, and some estimate that more than 50 million hectares could be developed in the infertile and unused Cerrado region of the central Brazilian Plateau.

The stem of the sugarcane plant grows 3–4 meters high. The stalks of sugarcane are harvested by hand or by harvester nine to fourteen months after planting. Only the stalks of sugarcane are removed from the field, and then they are sent to sugar mills. Sugarcane harvesting is labor-intensive work, and before harvesting, farmers often set fire to the dry leaves attached to the stalks. This decreases the quality of the sugar inside the stalk, but it avoids the process of removing the leaves. Because of air pollution, many governments are beginning to prohibit farmers from burning their harvests. Green harvests, including mechanical harvesting, can leave the tops, leaves, stubble, and roots as residues in the field. Thus, most of the nutrients in the residues are returned to the field and are absorbed by next crop of plants. In addition, some parts of plant nitrogen are derived from nitrogen fixation by endophytic nitrogen-fixing bacteria that live inside the plant tissues. The top of the sugarcane plant is removed at harvest and used as animal feed in some areas. Once planted, sugarcane can be harvested several times because new stalks, called ratoons, grow from the stubble.

Processing and By-Products

In the process of refining sugar at a sugar mill, bagasse, filter cake, boiler ash, and molasses are discharged as by-products. Bagasse is a fibrous residue left after squeezing sugarcane in a compressor; the main element of bagasse is cellulose. Most bagasse is burned as fuel for the sugar factory, which increases the energy self-sufficiency of the sugar mills. Excess bagasse is used as fuel for cogeneration to produce electricity and thermal energy. It is also used as pulp for paper or compressed to make boards for furniture.

Filter cake, or filter mud, is another processing residue. It includes minerals such as calcium, magnesium, nitrogen, phosphorus, potassium, and silicon. Therefore, filter cake is typically used as fertilizer or material for compost. Ash from the boilers, which contains minerals, also can be used as fertilizer. Molasses is a syrup and a final product that is separated from raw sugar during the manufacturing of crystallized sugar. The main elements of molasses are sucrose and reducing sugars, such as glucose and fructose. It is a raw material for the fermentation industry and can be further processed into monosodium glutamate or rum. Vinasse, which is the effluent residue from alcohol production, can be spread on fields as fertilizer.

Alternative Energy Uses

Bioethanol is a renewable energy resource that is produced from agricultural feedstock and blended with gasoline (gasohol) for use as a transportation fuel. In case of sugarcane, it is made from molasses or squeezed sugar juice. About the half of the sugarcane grown in Brazil is used to produce bioethanol. On average, a hectare-size field of sugarcane can produce 6 kiloliters of ethanol. (This assumes that 80 liters of ethanol can be produced from a metric ton of sugarcane stems, and that a field can yield 75 metric tons of sugarcane per hectare.)

Maize has been used for bioethanol production in the United States, and cassava is also other potential crop. But the starch of maize and cassava requires additional saccharization processing to turn it into sugar before alcoholic fermentation can take place. In addition, bioethanol production from sugarcane does not need an external energy supply because bagasse is used as fuel. Some research suggests that sugarcane is one of the most economically and environmentally advantageous crops for bioethanol production.

Shotaro ANDO
National Institute of Livestock and Grassland Science, Japan

See also in the *Berkshire Encyclopedia of Sustainability* Agriculture (*several articles*); Animal Husbandry; Bioenergy and Biofuels; Cacao; Coffee; Fertilizers; Food in History; Grasslands; Nitrogen; Tea

Further Reading

Alexander, Alex Getchell. (1985). *The energy cane alternative*. Amsterdam: Elsevier Science.

Bakker, Henk. (1999). *Sugar cane cultivation and management*. New York: Kluwer Academic / Plenum Publishers.

Better Sugar Cane Initiative Limited. (2010). BSI Public Consultation Standard Version 2. Retrieved July 13, 2010, from http://www.bonsucro.com/standard/assets/BSI%20Public%20Consultation%20Standard%20Version%202%2812%29a.pdf

Cheesman, Oliver D. (2004). *Environmental impacts of sugar production: The cultivation and processing of sugarcane and sugar beet*. Oxfordshire, UK: CABI Publishing.

Food and Agriculture Organization of the United Nations (FAO). (2010). *Crop production*. Retrieved May 11, 2010, from http://faostat.fao.org/site/567/default.aspx#ancor

James, Euan K. (2000). Nitrogen fixation in endophytic and associative symbiosis. *Field Crops Research*, *65*, 197–209.

James, Glyn. (Ed.). (2004). *Sugarcane* (2nd ed.). Oxford, UK: Blackwell Science.

Ponting, Clive. (2000). *World history: A new perspective*. London: Chatto & Windus.

União da Indústria de Cana-de-açúcar [Brazilian Sugarcane Industry Association]. (2010). *The virtual mill*. Retrieved July 13, 2010, from http://english.unica.com.br/virtual-mill/

Zuubier, Peter, & van de Vooren, Jos. (2008). *Sugarcane ethanol*. Wageningen, The Netherlands: Wageningen Academic Publishers.

Thorium

Recent expansion of the nuclear industry is depleting uranium reserves at unprecedented rates while rekindling worries about uranium's nuclear proliferation and waste risks. A greater reliance on thorium for nuclear energy could alleviate concerns about uranium's sustainability in coming decades, but its use must be managed sustainably as well.

Thorium (Th-232) is a radioactive element found throughout the Earth's crust, usually in diverse rock types containing uranium and rare earth metals. Its economically recoverable reserves are at least as abundant as uranium's, the traditional nuclear energy fuel, with some estimates claiming upward of three times the supply of uranium (OECD and IAEA 2010). The majority of world thorium deposits are in Australia, the United States, Turkey, India, and Venezuela (ThorEA 2010, 51).

Th-232 is nonfissile, meaning its atom cannot be split apart. It is fertile, however, which means its atoms can be made fissile; this is a crucial consideration, since fission is the crux of nuclear energy. Th-232 can be converted to fissile uranium-233 (U-233) using either a breeder reactor or an accelerator-driven subcritical reactor (ADSR). As the global nuclear power fleet has entered a period of aggressive expansion to quell climate change and provide energy security in an increasingly volatile energy environment, thorium's virtues make it a more attractive alternative to uranium.

Drivers of Thorium Growth

Developments around the world suggest increased thorium use in the near future. India signed a bilateral nuclear energy cooperation pact with the United States (the 123 Agreement) in 2009 that will greatly facilitate New Delhi's nuclear aspirations, much of which rely on thorium. Indeed, India has been the largest thorium-fueled civil nuclear power to date. While potentially beneficial to the sustainability of uranium, India's expansionary nuclear energy prospects raise nuclear proliferation concerns, as they rely on breeder reactors (WNA 2011).

In response to India's thorium use, Great Britain is pioneering a new technology, the accelerator-driven subcritical reactor (ADSR), which eliminates the need for a separate breeder reactor. ADSR breeds U-233 itself, burning plutonium as it is created, thus removing the risk of plutonium proliferation. Further, ADSR burns other waste products created during its fuel cycle that conventional thorium and uranium reactors must store for centuries. For these reasons, ADSR is seen by some as a solution not only for nuclear safety and waste considerations but uranium sustainability issues as well. But ADSR technology is still in its infancy: the first reactor is not scheduled to go online until 2025, although "demonstrators" will be developed and functioning by 2015 (ThorEA 2010, 7).

Thorium vs. Uranium

Th-232 is a potential solution to uranium's supply, waste, and nuclear proliferation concerns. Known thorium supplies are greater than those of uranium, which are projected to decline at an increasing rate due to an expanding nuclear industry. Under current practices, thorium creates less waste than uranium because virtually all mined Th-232 can be burned in a reactor; enrichment is thus unnecessary. This stands in contrast to the uranium fuel cycle, where it is common practice to isolate the 0.7 percent fissile portion of uranium ore (U-235) and discard the remaining 99.3 percent nonfissile portion (U-238). Not only must uranium ore undergo a time- and resource-consuming enrichment process to isolate the fractional amount of U-235, the discarded U-238 accumulates as waste (ThorEA 2010, 8). It should be noted, however, that U-238 is fertile and can be converted to fissile plutonium (P-239) in a breeder reactor, just as fertile Th-232 can be converted to fissile U-233. While both U-233 and P-239 can fuel nuclear weapons, the latter is more commonly used to this end. Thus, uranium's proliferation risk is said to be higher than thorium's.

ADSR technology eliminates nuclear proliferation risks because all of its fissionable fuel, as well as the fissionable by-products it creates, are burned in the reactor; in other words, unauthorized actors are unable to gain access to the fissionable fuel. Conventional thorium reactors, however, require a breeding process that creates large amounts of plutonium waste, which is a primary ingredient of atomic weapons.

Thorium Outlook

In addition to India and Britain's expansionary thorium prospects, Canadian, Chinese, French, and American companies in 2009 signed research and development agreements that could steer considerable investment toward the resource. According to the World Nuclear Association, thorium "holds considerable potential in the long-term. It is a significant factor in the long-term sustainability of nuclear energy" (WNA 2011).

Casey COOMBS
University of Utah

See also in the *Berkshire Encyclopedia of Sustainability* Mining—Metals; Uranium; Waste Management

Further Reading

Australian Atlas of Minerals Resources, Mines & Processing Centres. (2009). Thorium. Retrieved June 2, 2011, from http://www.australianminesatlas.gov.au/aimr/commodity/thorium_09.jsp

International Atomic Energy Agency (IAEA). (2005). *Thorium fuel cycle: Potential benefits and challenges*. Retrieved May 30, 2011, from http://www-pub.iaea.org/mtcd/publications/pdf/te_1450_web.pdf

Massachusetts Institute of Technology (MIT). (2009). *Update of the MIT 2003 future of nuclear power: An interdisciplinary MIT study*. Retrieved September 23, 2010, from http://web.mit.edu/nuclearpower/pdf/nuclearpower-update2009.pdf

Organisation for Economic Co-operation and Development (OECD) & International Atomic Energy Agency (IAEA). (2010). *Uranium 2009: Resources, production and demand*. Paris: OECD Publishing.

Thorium Energy Amplifier Association (ThorEA). (2010). *Towards an alternative nuclear future: Capturing thorium-fuelled ADSR energy technology for Britain*. Retrieved May 16, 2011, from http://www.thorea.org/publications/ThoreaReportFinal.pdf

World Nuclear Association (WNA). (2011). Thorium. Retrieved June 1, 2011, from http://www.world-nuclear.org/info/inf62.html

Uranium

Uranium is the primary fuel for atomic bombs and nuclear energy reactors. Both applications have altered the course of human history in profound ways. Nuclear energy production holds the potential to slow climate change and increase energy security worldwide. But if these global goals are to become a sustainable reality, finite uranium resources and nuclear safety risks—such as the March 2011 tsunami that crippled Japan's Fukushima Daiichi Nuclear Power Station—must be managed accordingly.

Uranium (U) is a radioactive metallic element found throughout the Earth's rocks, soils, and oceans, usually in low concentrations. Higher concentrations, which are economically feasible to extract, range from 0.01 percent U (very low-grade ore in Namibia) to 20 percent U (very high-grade ore in Canada) (WNA 2010c). Over half of the world's economically mineable uranium exists in and is produced by three countries: Canada, Australia, and Kazakhstan (OECD and IAEA 2010, 17, 45). Natural uranium consists of three isotopes in the following proportions: U-238 (99.284 percent), U-235 (0.711 percent), and U-234 (0.0055 percent). The element is perhaps best known for the energy-producing qualities of isotope U-235. Its nucleus can be fissioned, or split apart, to release tremendous amounts of energy, which can be channeled in nuclear reactors to power electrical grids across the globe. As of April 2011, 443 nuclear power plants produced approximately 14 percent of the world's electricity (WNA 2010b), although in the wake of the March 2011 tsunami that crippled Japan's Fukushima Daiichi Nuclear Power Station, the future of the nuclear power industry is uncertain: some countries (including Germany, Italy, and Japan) have already announced moratoriums on future nuclear power installations or closures of existing plants. Most countries, however, decided at an April 2011 meeting at the UN International Atomic Energy Agency (IAEA) headquarters in Vienna that they would delay such decisions pending safety reviews. Specifically, delegates from sixty-one of the seventy-two contracting parties to the Convention on Nuclear Safety (CNS), a nonbinding international nuclear safety treaty, pledged in Vienna to carry out safety reviews of existing nuclear installations, as well as reexamine the safety measures that guard against "extreme external events," such as tsunamis, in the future (IAEA 2011).

Uranium: Origins and Evolution

Uranium was discovered in the late eighteenth century. At the time, little was known about its potential to release energy in the form of a nuclear chain reaction. In subsequent years, the element came to be used for many purposes, including production of X-rays, as a multicolored ceramic glaze, and as photographic toner (Hammond 2000, 4–32). While those uses are still common, our understanding of uranium's capabilities changed dramatically in the years leading up to World War II.

In 1938, Italian, German, and French scientists demonstrated that the uranium atom was fissionable when bombarded with neutrons. The next year, scientists discovered that energy released from the initial fission could produce enough neutrons to split neighboring uranium nuclei, causing a self-sustained chain reaction. The result was unparalleled energy production.

Since its fissionable properties were discovered, uranium has been used primarily for two purposes: military weapons and civilian nuclear energy production. The first use, military arms, became evident to the world on

6 August 1945 when the United States dropped its first nuclear bomb on Hiroshima, Japan. Three days later, the second atomic bomb was dropped on neighboring Nagasaki. The nuclear warheads killed more than 200,000 of the towns' inhabitants—a level of destruction not witnessed again until the March 2011 tsunami that left upward of 25,000 people dead or missing, displaced hundreds of thousands, and put tens of thousands more at risk of radiation exposure from the stricken Fukushima nuclear reactors (GOJ 2011).

While only two nuclear weapons have ever been used in warfare, more than two thousand bombs have been tested and upward of twenty-six thousand built. Nearly all of this bomb making and testing took place while the Cold War waged between the West (the United States and its allies) and the East (the former Soviet Union and its allies), which lasted from the late 1940s until 1991. Two international treaties emerged from the Cold War arms race in an attempt to regulate nuclear weapons development and testing: the Nuclear Non-Proliferation Treaty (NPT), designed to limit the spread or proliferation of nuclear weapons to unauthorized individuals, and the Comprehensive Nuclear-Test-Ban Treaty (CTBT), designed to ban all nuclear explosions. Since 1998, six nuclear weapons have been tested by three countries—India, Pakistan, and North Korea—none of which is party to the NPT or the CTBT. Today, eight countries are known to have nuclear weapons: the United States, the Russian Federation, the United Kingdom, France, China, India, Pakistan, and North Korea. Israel is also widely assumed to have nuclear weapons (CTBTO 2010).

The other use of uranium's fissionable properties has been to generate electricity in civil or commercial nuclear power plants. Construction of nuclear power plants began in the mid-1950s and has spread throughout the world. At present, 443 commercial nuclear reactors operate in twenty-nine countries, 64 nuclear plants are under construction in fifteen countries, and over 150 more are scheduled to be built. Of the twenty-nine countries in which reactors operate, the United States has the largest number (104), followed by France (59) and Japan (54). Although the United States has far more nuclear reactors than any other country, its fleet only accounts for approximately 20 percent of the nation's total electricity production. A slightly larger portion of Japan's electricity comes from nuclear reactors (30 percent) and, until the 2011 tsunami, this was expected to grow to at least 40 percent by 2017. For the time being, however, nuclear power growth in Japan is uncertain pending safety reviews. Indeed, the Hamaoka nuclear power plant in central Japan was closed as of May 2011 until it is better protected against tsunamis. In contrast to the United States and Japan, nearly 80 percent of France's electricity is nuclear based. One reason that nuclear energy comprises a larger share of total electricity production in Japan and France is because they lack substantial reserves of domestic energy-producing resources, such as oil, natural gas, and coal. Another reason is that public opinion of nuclear safety in France and Japan has been less opposed to domestic nuclear industry growth than in the United States (IAEA 2010; NEI 2011; WNA 2010b). In light of Fukushima, however, public support for nuclear power in both Japan and France has waned, although the latter has not closed any plants.

Nuclear Fuel Cycle: Concepts and Concerns

Before uranium can produce electricity in nuclear reactors, it must go through several stages of preparation. Each stage must adhere to safety precautions to protect natural and human environments. First, uranium ore is mined from the Earth and transported to a mill, where it is ground, treated with chemicals, and purified into a powdery substance known as yellowcake (U_2O_8). Here, miners must protect themselves from inhalation and ingestion of carcinogenic radioactive dust and radon gas.

The yellowcake is then sold to various nuclear power companies that convert it to uranium hexafluoride (UF_6) for enrichment, a process that involves separating the highly fissionable isotope U-235 from its less fissionable counterpart U-238. Most nuclear reactors require uranium enriched to around 3–5 percent U-235. This enrichment level falls into the low-enriched uranium (LEU) category because it consists of less than 20 percent U-235. By contrast, nuclear warheads require enrichment levels of at least 85 percent; this enrichment level falls into the highly-enriched uranium (HEU) category. At the enrichment stage, all yellowcake purchased by a given nuclear company must be enriched to the precise levels specified by the reactors it will fuel as a cautionary measure to prevent proliferation of HEU, the primary ingredient in nuclear weapons.

Once enriched to required low levels, the uranium is used to fabricate reactor fuel rods, which generate electricity in the reactor core. At the service stage, when the fuel rods are generating electricity, safety measures must be taken to prevent accidents such as reactor meltdowns and radiation leaks. In over fourteen thousand cumulative years of commercial reactor operation in thirty-two countries, few major service-stage accidents have occurred, although three in particular stand out: Three Mile Island, Chernobyl, and Fukushima. The steps discussed up to this point represent the "front-end" of the nuclear fuel cycle.

The "back-end" of the nuclear fuel cycle consists of containment, reprocessing, or disposal of the fuel cycle's waste products, which include waste rock from mining, mill tailings, depleted uranium (i.e., leftover uranium from the enrichment process), and spent or used nuclear fuel. The back-end of the cycle presents unique environmental and human safety risks, as weapons-grade plutonium is created from burning uranium fuel rods; this stage is arguably most vulnerable to nuclear proliferation.

Two types of nuclear fuel cycles exist: open and closed. A fuel cycle is said to be open if depleted uranium and spent fuel are not recycled. "Once-through" reactors have an open fuel cycle because they do not reuse the fissionable waste created from enriching and burning fuel. Instead, the waste is prepared for long-term storage. It should be noted that no permanent nuclear waste storage facility (such as the one proposed at Yucca Mountain, Nevada) are available. For the time being, most waste worldwide is housed either at the nuclear power plant from which it came or shipped to temporary storage sites. In contrast to the open cycle, a closed fuel cycle recycles waste products such as spent fuel and depleted uranium. "Breeder" reactors, which produce more fissile material (e.g., plutonium 239) than they burn, operate on a closed fuel cycle.

Changing Landscape of Uranium Mining

Historically, the two most common uranium extraction methods have been open-pit and underground mining. Both methods require extracting large amounts of ore from the Earth. Since only a fraction of mined ore is uranium (generally 0.01–20 percent), and fissile U-235 makes up only 0.07 percent of this uranium, considerable amounts of waste rock and mill tailings accumulate. These piles of waste tailings must be transported to a repository that will not only contain radioactive dust, radon gas, and arsenic, but will be able to withstand floods, landslides, earthquakes, and tsunamis for hundreds to thousands of years (Diehl 2004). A number of inadequately contained sites, known as legacy sites, currently exist throughout the world due to insufficient regulatory systems or safety standards that have existed for the majority of uranium mining's history (IAEA 2007). Many of these sites are now being remediated or cleaned up using public funds, such as the $200 million dollar Moab Mill Tailings Cleanup Project in Utah (DOE 2010, 1). Some countries are attempting to prevent creation of new legacy sites by mandating environmental assessment processes prior to mine openings or expansions (e.g., Australia, Canada), implementing monitoring programs at mines already in operation (e.g., Kazakhstan), facilitating efforts to reduce water consumption (e.g., Namibia), and establishing more stringent environmental radiation protection regimes (e.g., China) (OECD and IAEA 2010, 11).

Within the last five years, in situ leaching (ISL) has begun to emerge as a dominant mining method. ISL differs greatly from the two traditional methods of mining. Instead of removing ore from the Earth and sending it to a mill, a leaching liquid is injected into ore deposits via a system of wells, and the separated uranium is pumped out. While ISL does not create large amounts of potentially harmful waste rock and mill tailings, it does leave residual leaching liquid in the ore deposits. Precautions must be taken to ensure that leaching liquids do not contaminate underground water supplies. Although it is currently only used in sandstone formations, the method is expected to overtake both open-pit and underground mining as a percentage of world production. Most of the recent increases in ISL activity have taken place in Kazakhstan, Australia, China, Russia, the United States, and Uzbekistan (OECD and IAEA 2010, 52–53).

Three Waves of Uranium Production

Three periods of significant world uranium production have occurred since the1940s, all of which were influenced by political, economic, and social considerations.

The initial period of intensive uranium extraction lasted from the mid-1940s to the mid-1960s. The uranium spot price (i.e., the price quoted for immediate payment and delivery of yellowcake) peaked in 1953 at approximately $200 per kilogram of uranium (kgU) and worldwide production peaked in 1959 at approximately

50,000 tons (Neff 1984, 88). This sustained period of mining was propelled by a nuclear arms race between the United States and the former Soviet Union. In the United States, the Atomic Energy Commission (AEC), established by President Truman on 28 October 1946, was the sole buyer of uranium. The AEC provided incentives and subsidies to encourage extraction (Leach 1948). Uranium mining in the United States during the twenty-year period centered on the Colorado Plateau.

During roughly the same period, the Soviet Union extracted uranium from mines in the Ore Mountains, which straddle eastern Germany and the Czech Republic. As the sole buyer of uranium from these mines, the USSR used various methods to stimulate production, including forced labor from convicts and monetary incentives for the willing (Heitschmidt 2003).

Commercial nuclear energy programs also began to grow during this period, but their demand for uranium developed more slowly than expected. Nations around the world, particularly the United States and the USSR, began to accumulate sufficient stockpiles of nuclear weapons and demand less uranium. The outcome of decreased military demand and delayed civil demand was a gross oversupply of the fissionable resource, which drove its selling price and production down.

The second period of uranium extraction lasted from the mid-1970s to the late 1980s, with spot price peaking from 1975 to 1977 at approximately $275/kgU, and worldwide production cresting in 1979 at approximately 70,000 tons. The increase in production was driven mainly by the maturation of civil nuclear energy programs and the 1970s oil crisis, the latter of which encouraged political leaders to consider alternative sources of energy. High annual production levels (60,000–65,000 tons) continued from the late 1970s to the late 1980s before experiencing a sharp decline to 30,000–40,000 tons per year, which became the norm until 2005 (OECD and IAEA 2010, 87).

Two disasters involving nuclear reactors triggered the slow decline of uranium production during its second wave. The first incident occurred in March 1979 when a partial reactor meltdown at the Three Mile Island (TMI) nuclear power plant in Harrisburg, Pennsylvania, released radioactive gas into the atmosphere. According to independent studies since 1979, no significant rise in cancer rates around Three Mile Island have been directly linked to the meltdown (Kemeny 1979; Talbott et al. 2003). The TMI disaster, in combination with delayed construction schedules and overrun costs, was largely responsible for the more than thirty-year ban on new plant construction in the United States starting in the late 1970s (MIT 2009, 8). The second disaster occurred in April 1986 when the Chernobyl nuclear reactor in what is now Ukraine exploded. Atmospheric radioactive debris from the accident extended throughout eastern, northern, and western Europe. At least sixty people died from the explosion, around fifty have already died from illnesses related to radiation exposure, and as many as four thousand people may eventually die from radiation-related illnesses (IAEA, WHO, and UNDP 2005). Shortly after Chernobyl, global production and the spot price of uranium plummeted due primarily to public fear of the power source.

A similar, though less severe, drop in spot price was observed in the immediate aftermath of Fukushima. The price has since begun to rise to predisaster levels and is expected to continue to rise for two reasons: current world uranium supply remains well short of demand, and the majority of future demand will come from state-run countries, such as China, where public input is bypassed (Matta 2011).

TMI, Chernobyl, and Fukushima brought public awareness to the potential dangers of nuclear energy. As a result, public opinion has influenced political decisions on numerous occasions regarding continued nuclear power plant construction. That, in turn, has affected uranium production levels because mining companies are less willing to allocate limited resources to the exploration and extraction of the natural resource when its future demand is so vulnerable to public opinion and the political process.

The third and most recent wave of uranium production started around 2005 when its spot price began to climb in response to four global variables affecting supply and demand. First, India, China, and South Korea announced plans to build a significant number of new reactors. Globally, China is leading production, with nearly half of all active construction projects in 2009, and is projected to add the most nuclear capacity through 2035 (US EIA 2010, 80). Second, momentum began to build in the

United States to resume expansion of its nuclear fleet (NEA 2009). It should be noted that the new growth prospects in these four countries are indicative of growing worldwide concerns about energy security and reducing greenhouse gas emissions (GHG). Third, a large percentage of the world's existing nuclear fleet was ready to be refueled. Fourth and finally, all of the preceding demand-side factors were exacerbated when two of the world's most productive mines—Cigar Lake Mine in Canada and Ranger Mine in Australia—flooded, creating short-term fears of undersupply. The result was a 2007 spike in spot price to $354/kgU (though the price stabilized to $116/kgU at the end of 2008), which stimulated new exploration and intensified existing mining activities. In 2009 production levels exceeded 50,000 tons for the first time in twenty years and are expected to continue to grow, although perhaps at a slower rate in light of the Fukushima disaster, to support an expanding nuclear fleet (OECD and IAEA 2010, 97).

Sustainable Options?

Since intensive uranium mining began during the 1940s and 1950s, it is estimated that 2,415,000 tonnes of uranium (tU) have been extracted. This represents just under half of identified world reserves—5,404,000 (tU)—economically recoverable at a spot price of $130/kgU. At 2008 rates of uranium consumption by the global nuclear power fleet (59,065 tU), tU reserves are expected to last for over one hundred years. By 2035, however, annual uranium requirements for the world's growing fleet are projected to range from 87,370 tU to 138,165 tU, which would deplete current known reserves decades sooner than expected (OECD and IAEA 2010, 10).

In the face of an expanding nuclear industry, several options have emerged to budget the world's uranium supply. One is to increase supply by identifying the estimated 10,400,000 tons of undiscovered uranium in the Earth's crust. Undiscovered resources are defined by OECD/IAEA (2010, 27) as "resources that are expected to occur based on geological knowledge of previously discovered deposits and regional geological mapping." It is presently unclear whether such reserves would be economically feasible to extract. Another possibility is to develop economically sound technologies to extract commercial-scale quantities of uranium from the ocean, whose supply of 4 billion tons is virtually inexhaustible. For the foreseeable future, however, that is not possible (OECD and IAEA 2010, 11, 32).

A third possibility is to use thorium as nuclear fuel in one of two ways. The first involves burning nonfissile natural thorium (Th-232) in a breeder reactor to produce fissile U-233. The U-233 can then be used to fabricate fuel rods for closed-fuel-cycle reactors. The second way to burn thorium as fuel in a nuclear reactor is through accelerator-driven sub-critical reactor (ADSR) technology, which does not require a breeder reactor to convert Th-232 to U-233. Thorium use is a potentially viable way to conserve global uranium stocks for two reasons. First, mineable thorium reserves approximate those of uranium (OECD and IAEA 2010, 32–33); and second, thorium enrichment is unnecessary because virtually all of the mined Th-232 can be used in a reactor, whereas only 0.7 percent of U-235 in natural uranium is usable. The first technology, using a breeder reactor to produce U-233, has yet to become popular as a substantial source of nuclear energy due to both the high costs of fabricating U-233 rods from Th-232 and the nuclear weapons proliferation risks that arise from this fabrication process. Currently, it is only being used on a large scale in India. ADSR technology, by contrast, is gaining attention as a nuclear fuel source in Britain. The first ADSR reactor is scheduled to become operational by 2025 (ThorEA 2009–2010).

At present, the most widely used option to budget uranium supplies is to recycle waste from the nuclear fuel cycle—specifically, spent nuclear fuel and depleted uranium. Spent nuclear fuel can be recycled in two ways. The first is to salvage weapons-grade plutonium (P-239) created during the initial fission process and blend it with uranium. This P-239/U blend, known mixed-oxide fuel (MOX), requires specialized reprocessing and fuel fabrication facilities and is currently used in 6 percent of the world's operating fleet. It is estimated that MOX fuel saved 1,972 tU, or 4.5 percent of world uranium production, in 2008 (OECD and IAEA 2010, 92–93). The second way to recycle spent fuel entails reprocessing leftover uranium that has passed through a reactor, since 96 percent of it is still fissionable. Due to the high costs inherent in conversion, enrichment, and fabrication facilities, reprocessed uranium (RepU) is used marginally as a recycled fuel, accounting for less than 1 percent of 2009 uranium production. At present, the United Kingdom, France, and Russia are the only countries that routinely reprocess uranium (OECD and IAEA 2010, 92–94).

The other recyclable waste product of the nuclear fuel cycle is depleted uranium, which is created during the initial uranium enrichment process. Once reenriched, depleted uranium can be used in place of the standard 3–5 percent LEU. Current global inventory of depleted uranium, estimated to be around 1,600,000 tU, would produce about 450,000 tU of equivalent natural uranium. This is enough to operate the global nuclear fleet at 2006 levels for seven years (NEA 2007). Yet reenrichment of depleted uranium is practiced minimally throughout the world, since it is economically feasible only in certain enrichment facilities with extra capacity and low operating costs.

The debate over whether to use breeder or once-through reactors is intense and ongoing. Proponents of breeder technology claim that it uses uranium sixty to one hundred times more efficiently than once-through reactors (WNA 2010a; ANS 2005, 1). Others routinely note the prohibitive costs of breeders and the fact that they are specifically designed to create weapons-grade plutonium (or weapons-grade U-233 from breeding Th-232) to be reused in the reactor. This last point is the nuclear proliferation risk cited frequently by critics of closed-cycle breeder reactors. The United States, and most other countries with civil nuclear reactors, has opted for once-through technology because it is less expensive and reduces the risk of unauthorized individuals acquiring P-239. Yet research and development funds are being devoted to developing safer and cheaper closed fuel cycles, with the hope that one day sound economics and a sufficient security apparatus can coexist (NAS 2008, 56; MIT 2009, 16).

Future Prospects

As of 2010, the US Energy Information Administration (EIA), Organization for Economic Co-operation and Development (OECD), and IAEA expected strong growth in the nuclear industry as early as 2015, perhaps doubling in capacity by 2035 (US EIA 2010, 80; OECD and IAEA 2010, 102). Although post-Fukushima growth projections will be revised following worldwide safety reviews of existing installations and reexamination of safety measures that guard against "extreme external events," such as earthquakes and tsunamis, rising uranium demand from rapidly developing countries should ensure its continued expansion.

Uranium holds the potential to end wars, start wars, bring electricity to millions, and mobilize millions in opposition to its use as a growing energy source. If nuclear power is going to play a sustainable role in reducing GHGs to curtail climate change and bringing energy security to countries the world over, uranium reserves and nuclear fuel cycle risks must be managed accordingly. Though certainly not an easy task, carving out an appropriate role for nuclear fission in the twenty-first century and beyond is achievable. Willingness from all sides of the debate to make concessions and learn from each other is paramount.

Casey L. COOMBS
University of Utah

See also in the *Berkshire Encyclopedia of Sustainability* Coal; Mining—Metals; Solar Energy; Thorium; Water Energy; Wind Energy

Further Reading

American Nuclear Society (ANS). (2005). Fast reactor technology: A path to long-term energy sustainability. Retrieved September 24, 2010, from http://www.ans.org/pi/ps/docs/ps74.pdf

Comprehensive Nuclear-Test-Ban Treaty Organization (CTBTO). (2010). Nuclear testing. Retrieved October 8, 2010, from http://www.ctbto.org/nuclear-testing/

Department of Energy (DOE). (2010). Audit of Moab Mill tailings cleanup project. Retrieved September 24, 2010, from http://www.ig.energy.gov/images/OAS-RA-L-10-03.pdf

Diehl, Peter. (2004). Uranium mining and milling wastes: An introduction. Retrieved August 30, 2010, from http://www.wise-uranium.org/uwai.html

Government of Japan (GOJ). (2011). Current situation and the Government of Japan's response, 6 May 2011. Retrieved May 10, 2011, from http://reliefweb.int/node/400601

Hammond, C. R. (2000). Handbook of chemistry and physics: The elements. New York: CRC Press.

Heitschmidt, Traci. (2003). The quest for uranium: The Soviet uranium mining industry in eastern Germany, 1945–1967. Santa Barbara: University of California.

Institute for Energy and Environmental Research (IEER). (2005). Uranium: Its uses and hazards. Retrieved September 1, 2010, from http://www.ieer.org/fctsheet/uranium.html

International Atomic Energy Agency (IAEA). (2007). Uranium mining legacy sites and remediation: A global perspective. Retrieved October 11, 2010, from http://www.iaea.org/OurWork/ST/NE/NEFW/documents/RawMaterials/CD_TM_Swakopmund%20200710/13%20Waggit4.PDF

International Atomic Energy Agency (IAEA). (2010). Power reactor information system (PRIS). Retrieved October 11, 2010, from http://www.iaea.org/programmes/a2/index.html

International Atomic Energy Agency (IAEA), World Health Organization (WHO), & United Nations Development Programme (UNDP). (2005). Press release: Chernobyl: The true scale of the accident. Retrieved January 6, 2011, from http://www.iaea.org/newscenter/focus/chernobyl/pdfs/pr.pdf

International Atomic Energy Agency (IAEA). (2011). Nuclear Safety Convention meeting commits to learn lessons from Fukushima nuclear accident. Retrieved May 6, 2011, from http://www.iaea.org/newscenter/news/2011/cnsmeetingends.html

Kemeny, J. G. (1979). Report of the President's Commission on the accident at Three Mile Island: The need for change: The legacy of TMI. Retrieved January 6, 2011, from http://www.threemileisland.org/downloads/188.pdf

Leach, Paul, Jr. (1948). Uranium ore: How to go about finding and mining it. *Engineering and Mining Journal, 149*(9), 75–77.

Matta, Jaya. (2011). DJ interview: Uranium energy: Prices to rally despite Japan crisis. Retrieved May 10, 2011, from http://www.morningstar.co.uk/uk/markets/newsfeeditem.aspx?id=138501958094799

Massachusetts Institute of Technology (MIT). (2009). Update of the MIT 2003 future of nuclear power: An interdisciplinary MIT study. Retrieved September 23, 2010, from http://web.mit.edu/nuclearpower/pdf/nuclearpower-update2009.pdf

National Academy of Sciences (NAS). (2008). Review of DOE's nuclear energy research and
development program. Retrieved September 4, 2010, from http://books.nap.edu/openbook.php?record_id=11998&page=57

Neff, Thomas L. (1984). The international uranium market. Cambridge, MA: Ballinger Publishing.

Nuclear Energy Agency (NEA). (2007). Management of recyclable fissile and fertile materials. Paris: Organisation for Economic Co-operation and Development.

Nuclear Energy Agency (NEA). (2009). Nuclear energy data. Paris: Organisation for Economic Co-operation and Development.

Nuclear Energy Institute (NEI). (2011). World statistics: Nuclear energy around the world. Retrieved May 10, 2011, from http://www.nei.org/resourcesandstats/nuclear_statistics/worldstatistics/

Organisation for Economic Co-operation and Development (OECD) & International Atomic Energy Agency (IAEA). (2010). Uranium 2009: Resources, production and demand. Paris: Organisation for Economic Co-operation and Development.

ReliefWeb. (2011). Countries + disasters: In-depth profiles, updates and reports on countries and disasters: Japan tsunami. Retrieved May 10, 2011, from http://reliefweb.int/taxonomy/term/128

Ringholz, Raye C. (2002). Uranium frenzy: Saga of the nuclear West. Logan: Utah State University Press.

Talbott, Evelyn O.; Youk, Ada O.; McHugh-Pemu, Kathleen P.; & Zborowski, Jeanne V. (2003). Long-term follow-up of the residents of the Three Mile Island accident area: 1979–1998. *Environmental Health Perspectives, 111*(3), 341-348.

Thorium Energy Amplifier Association (ThorEA). (2009–2010). Towards an alternative nuclear future: Capturing thorium-fuelled ADSR energy technology for Britain. Retrieved May 16, 2011, from http://www.thorea.org/publications/ThoreaReportFinal.pdf

US Energy Information Administration (US EIA). (2010). International energy outlook 2010. Retrieved September 3, 2010, from www.eia.gov/oiaf/ieo/index.html

US Environmental Protection Agency (EPA). (2002). Facts about uranium. Retrieved September 15, 2010, from http://www.epa.gov/superfund/health/contaminants/radiation/pdfs/uranium.pdf

World Information Service on Energy (WISE). (2006). WISE uranium project: Uranium mining and milling. Retrieved August 28, 2010, from http://www.wise-uranium.org/stk.html?src=stkd01e

World Nuclear Association (WNA). (2010a). Fast neutron reactors. Retrieved September 22, 2010, from http://www.world-nuclear.org/info/inf98.html

World Nuclear Association (WNA). (2010b). Nuclear power in the world today. Retrieved September 11, 2010, from http://www.world-nuclear.org/info/inf01.html

World Nuclear Association (WNA). (2010c). Supply of uranium. Retrieved September 11, 2010, from http://www.world-nuclear.org/info/inf75.htm

Zoellner, Tom. (2009). Uranium: War, energy, and the rock that shaped the world. London: Viking Penguin.

Utilities Regulation

The traditional system of public utility regulation in the United States and other nations has transformed dramatically since the 1970s. New laws and policies to encourage renewables, conservation, and energy efficiency programs have changed the electric industry from its traditional focus on increasing production to incorporating more environmental values. Much more is still necessary for electric utilities to become environmentally friendly and demand responsive.

Until recently, regulation of electricity production in the United States and elsewhere did not reflect its full social costs for two fundamental reasons. First, pollution caused by electric utilities was unregulated, allowing utilities to avoid the costs of controlling it. Since the 1970s, the system of air pollution regulation, beginning in the United States with the 1970 Clean Air Act Amendments and adopted elsewhere, has changed this substantially. Second, under traditional utility regulation in the United States, utilities had no incentive to adopt energy efficiency and demand response (DR) programs, such as time-based pricing, to reduce consumption. Under traditional regulation, utilities have a strong incentive to increase sales, because their costs are largely fixed. Increased sales mean increased profits, thus there is no incentive to reduce sales and profits. DR and energy efficiency programs reduce utilities' sales, and utilities resisted them as restaurants would shun diet plans. Yet these programs have enormous potential. Energy efficiency and DR saved utilities 32,741 megawatt hours in 2008 (US Department of Energy 2010), and much more is possible.

Traditional regulation did not give utilities any incentive to purchase power generated by solar, wind, and other renewable energy facilities. As monopolies, utilities were the only potential buyers for this power (Eisen 2010b), and they refused to deal with companies they saw as potential competitors. They viewed renewable power as more expensive than electricity generated from fossil fuels; however, the cost of electricity made from coal, oil, and natural gas did not reflect its full environmental costs (Herzog, Lipman, Edwards, and Kammen 2001). Utilities also pointed to other purported disadvantages of renewables (e.g., wind power's intermittent nature) as reasons to avoid purchasing it (Eisen 2010b).

The relationship between environmental values and utility regulation has changed dramatically since the 1970s, as regulatory changes have attempted to internalize the full costs of generating and transmitting electricity by promoting energy efficiency, DR, and renewables. Unfortunately, the "restructuring" of the 1990s and early 2000s, a general term for deregulatory initiatives that modified or eliminated traditional regulation systems, had a profoundly negative impact on utilities' environmental programs. Today, a wide variety of state and federal initiatives promote conservation, efficiency, and renewables.

Shortcomings of Traditional Utility Regulation

Public utilities such as electric and gas companies provide essential services to the public. Economists describe these companies as having natural monopoly positions, because the large amount of capital required to build infrastructure creates formidable barriers for new companies to compete with existing utilities. If left unregulated, utilities could exercise their monopoly positions to set above-market prices. The electric utility regulatory system is complex; electricity is often produced in one state and distributed in another through the transmission grid. Generally speaking, state public utility commissions (PUCs) regulate local utility operations, including rate setting, and the Federal Energy Regulatory Commission (FERC) regulates interstate wholesale of electricity and interstate transmission grid operations.

The first decades of utility regulation focused on curbing utilities' monopolistic tendencies through price regulation. In this system, state PUCs granted utilities valuable franchises, giving them sole rights to serve all customers in a specific territory. In return, utilities accepted state regulation of rates, known as cost of service (COS) regulation because it was based on calculating the utilities' costs of providing service, plus a fair rate of return. Understanding this process is critical to understanding how environmental values were traditionally disregarded. Regulators begin by setting the revenue the utility must recoup to meet its costs. PUCs make judgments about power plant, fuel, capital, and operating costs based on hearings in which utilities and other interested parties present testimony. Rates are typically fixed for several years until the next rate case, but utilities can recover intermediate cost increases with fuel adjustment clauses. Once required revenue is established, the PUC sets rates; for example, it might set the price per kilowatt hour (kWh) as revenue divided by kWh sold.

COS regulation does not give utilities any incentive to minimize costs. They have what economists call the "throughput incentive": all the costs they incur are passed through to consumers and recovered from them. In a famous paper, the economists Harvey Averch and Leland Johnson demonstrated that this gives utilities incentives to increase capital expenditures and other costs without limit, making them inefficient (Averch and Johnson 1962). Under COS regulation, utilities have every incentive to sell as much electricity as they can. Their costs are largely fixed, so the more they sell, the more revenue they generate. Political scientists argued that over time, regulators failed to reverse these incentives and force utilities to cut costs because utilities gained excessive influence over regulators, "capturing" them (Estache and Martimort 1999).

Until the 1970s, utilities expanded virtually without limit. The industry's mascot, Reddy Kilowatt, boasted that "things work better electrically" and promoted consumption to millions of Americans. New products using electricity were designed and marketed, such as the "all electric kitchen." Increasing demand led to the construction of new power plants, and PUCs aided in this expansion process by keeping electric rates low.

Rate Regulation Evolves

In the 1970s, electricity rates began to increase, and consumers started to object to utilities' unchecked growth. A primary driver of rate increases was rapid cost escalation in construction of new nuclear power plants, which coincided with antinuclear opposition from environmental groups. When crises in the Middle East threatened the supply of imported petroleum, consumers pressured utilities to consider conservation techniques. President Jimmy Carter pushed for a national energy policy, setting an example by keeping the heat low and wearing a sweater in the White House, and telling Americans that conserving energy and weaning the nation off foreign oil was the "moral equivalent of war."

Responding to these societal forces, the utility rate-setting process began to evolve. The Public Utility Regulatory Policies Act of 1978 (PURPA) was one of the laws that responded to the 1970s energy crises. PURPA revised rate-making structures and directed states to provide incentives encouraging electric utilities to establish conservation programs. Over time, these changes had a strong influence on utilities. In 2000, 962 American electric utilities had one or more efficiency and DR programs, and the 516 largest utilities saved 53.7 million kWh of electricity (US Department of Energy 2002).

To promote conservation by utilities themselves, states enacted statutes requiring utilities to consider various options to meet increasing demand in a planning process called integrated resource planning (IRP), which originated in the 1970s in California. A utility implementing IRP forecasts a range of future demand possibilities and considers all alternatives to meet that demand, including conservation and renewables (Cavanagh 1986). IRP is different from traditional utility planning, in which utilities sought to meet demand with the construction of new power plants and gave little consideration to conservation or renewables. The Energy Policy Act of 1992 directed states to consider IRP, and many did so.

PURPA's Avoided-Cost Mandate

In the 1970s, traditional electric utilities provided virtually all new generating capacity, incorporating little generation from renewables. They generally refused to purchase power from nonutility companies. Changing the standards of business, PURPA required utilities to purchase power from cogenerators and other small power producers. Rates were set at utilities' avoided cost—the cost of generating power, had the utilities not purchased it elsewhere. This statutory provision and the regulations implementing it had a revolutionary impact on small power production, especially in states such as California that set relatively high avoided-cost rates. At the same time, technologies matured, and renewable resources, such as solar and wind power, experienced dramatic growth.

The avoided-cost mandate had an important side effect. Until the 1990s, investor-owned utilities that were vertically integrated, performing all functions of generating, transmitting, and distributing electricity in their service territories, generated two-thirds of the United States' electric power. PURPA led to the rise of an entire class of nonutility generators (NUGs), merchant firms that generate electricity and compete with existing utilities but are not utilities themselves (e.g., they do not own transmission and distribution facilities).

Restructuring Transforms the Industry

The introduction of competition through PURPA was one of many trends that led to calls for utility deregulation in the early 1990s. Free market economists had achieved deregulation of the airline industry in 1978, and other industries were transforming as well. Many believed the largely successful restructuring already underway in the natural gas industry could serve as a model for deregulating electric utilities and providing electricity to consumers at a lower cost (US Department of Energy 2002). The FERC spurred restructuring with a series of orders that called for open access to the nation's transmission grid as well as regional transmission organizations to control the transmission grid, transforming the industry by operating new wholesale electricity marketplaces. The FERC required that utilities separate their transmission divisions from their generation and distribution divisions. In response, utilities began to break up into separate businesses.

In addition, about twenty-five states introduced retail choice, giving consumers the ability to choose their own electric suppliers. This choice, however, turned out to be largely illusory. In California, retail choice was a spectacular failure for numerous reasons, including market manipulation by Enron and others. The failure of restructuring in California made other states cautious, and few states experienced much, if any, retail competition between new companies and incumbent utilities. There were many reasons for this, including state statutes that kept electricity rates low, making market entry difficult. Today, competition in the retail market is largely discredited, and many states have abandoned restructuring and returned to traditional regulation. FERC continues to regulate wholesale electricity sales with regulations that address excess market power.

Impact on Environmental Programs

A central debate during utility restructuring was the idea of "stranded costs." Utilities had spent money, planning to recoup it over the following decades. They argued that states were forcing them to compete immediately with new generators who did not have to incur the same costs. In response, states allowed existing utilities to recover these costs from customers, and to discontinue programs that made it uneconomical for them to compete with new generators. This policy choice had major negative impacts on existing environmental programs, which utilities thought of as burdensome in a competitive environment (Black and Pierce 1993). During the transition to competition, these were among the first programs to be cut, and utilities' budgets for conservation and DR programs fell by over 50 percent nationwide (Brown and Sedano 2003). IRP also became less common, because state PUCs had less control over utilities' supply planning decisions in a competitive environment (DSIRE 1998).

Many also advocated repeal of the avoided-cost mandate. Ironically, PURPA's success helped bolster this position. There were far more NUGs in the mid-1990s than in 1978, and many believed that in a more competitive industry, NUGs should compete equally with other electricity suppliers (Black and Pierce 1993). Specific industry conditions exacerbated the problem. Utilities were obligated to continue purchasing power under long-term contracts with small power producers, even though generation costs had dropped. This, they argued, was a form of stranded costs that kept electricity rates high. In 2005, Congress responded by discontinuing the avoided-cost requirement in areas of the United States served by wholesale electricity markets.

System Benefits Charges

As efficiency and DR programs were being discontinued, a number of states stepped into the gap and established statewide charges, imposing a small fee on all customers' electric bills. While the fee imposed is typically small, the funds raised can amount to millions of dollars. States use these system benefit charges for purposes perceived to benefit the entire public, such as efficiency and DR programs, or rebates on renewable energy systems. The programs vary considerably from state to state, and are not a substitute for more comprehensive action to promote conservation, DR, and renewables.

Renewable Electricity Standards

To promote renewables, states have established renewable portfolio standards or renewable electricity standards (RES); the terms are interchangeable. A RES typically requires a utility to obtain a specified percentage of the power it sells from renewable sources, or to compensate by purchasing tradable credits from suppliers that have excess production. The RES is meant to encourage renewable energy deployment, stimulate industry growth, and make renewable power more competitive in price. In 2010, thirty-five states had RES mandates, voluntary goals for renewable power generation, or other RES-like programs (Pew Center on Global Climate Change 2010).

The American Clean Energy and Security Act climate bill, passed by the US House of Representatives in 2009, would have established a national RES, including both energy efficiency and electricity produced from renewables. Some believe that replacing state standards with a single, national RES and credit marketplace would provide more support for renewables (Davies 2010). While many in Congress support a national RES, others have blocked progress, citing potential for higher consumer costs and unequal regional impacts.

Other nations have had success with national targets for renewables. In 2008, the EU Renewable Energy Directive established a binding target of a 20 percent share of renewable energy sources in energy consumption by 2020. All twenty-seven European nations have some form of regulatory mandate or financial incentive designed to meet their individual national targets for renewables (Commission of the European Communities 2008). China has also made considerable progress in increasing use of renewable energy systems. While 80 percent of the nation's electricity is still generated from coal (Eisen 2010a), China has a strong Renewable Energy Law and aggressive national targets.

Feed-In Tariffs

One frequent objection to RES is that they require building more renewable energy facilities. To critics, this forecloses any possibility that other policies can do the same job at a lower cost. A different idea for promoting renewables currently gaining attention is the feed-in tariff (FIT). Under the FIT, renewable energy project owners are paid an above-market rate. The rate is locked in for a specific term of years, and is either a fixed amount defined in advance, or a premium over the wholesale price of electricity. This is in contrast to a RES because it focuses on the generator's cost and profit, not the utility's supply procurement process. The direct payment aims to generate a reasonable profit and make project financing easier; many support it because they believe it will lead to more deployment of renewables (Rickerson, Bennhold, and Bradbury 2008). Others, however, view paying above-market rates to renewable energy producers as anticompetitive (Eisen 2010b).

The FIT is based on successful European programs. The German FIT caused explosive growth in solar and wind power, and enabled Germany to more than double its supply of electricity produced from renewables between 2000 and 2007 (Rickerson, Bennhold, and Bradbury 2008). Other EU countries and the Canadian province of Ontario have adopted FITs. A number of US states have considered FITs, but by 2009, only three states and some localities had adopted them.

There is an overlap between a FIT and net metering initiatives underway in many states. Net metering rewards a customer for installing a renewable energy system that sometimes provides more than enough electricity to meet the customer's needs, and the customer sells power to the grid. FITs, by comparison, can pay anyone for electricity generated from renewables, regardless of a specific customer's needs. States need to design FITs carefully so that they work with net metering programs.

Real-Time Pricing and Green Pricing

Some believe that programs for real-time pricing (RTP) of electricity for residential utility customers have enormous promise to help reduce the demand for electricity and yield billions of dollars in savings for consumers. In 2008, however, only 1.1 percent of all US customers were enrolled in these programs (US Department of Energy 2009). In its simplest form, RTP involves providing price signals to consumers in real time, allowing them to adjust their behavior by cutting back on demand when prices are high. This would require wide adoption of a form of technology that is uncommon today. The standard electric meter gives no information to consumers about the price of electricity, thus more advanced meters are necessary for RTP be successful.

Green pricing programs also promote electricity sources that do not rely on fossil fuels. More than five hundred utilities offer these voluntary programs, which allow customers to purchase electricity based on its fuel source and emissions profile. These programs are limited in effectiveness for two primary reasons. First, they often require consumers to pay a price premium for renewable energy, which many consumers are unwilling to do. Second, because these programs are voluntary, they do not have large numbers of consumers enrolled. Therefore, green pricing is not likely to be as successful as specific, targeted mandates in encouraging the development of renewables.

Increased Deployment of Renewables

Increased deployment of renewables will require substantial improvements to the United States' transmission grid, as there has been chronic underinvestment in the electricity transmission and distribution system. Hundreds of thousands of high-voltage transmission lines cross the nation, but only 668 additional miles have been added since 2000 (US Department of Energy 2008). Many potential wind energy projects cannot connect to the grid, due to a lack of transmission lines in remote locations, where the wind tends to blow more strongly. Inadequate transmission capacity is also a problem in nations such as China, where solar and wind projects are located far from eastern urban centers. China expects to build many new ultra-high-voltage transmission lines in the next decade.

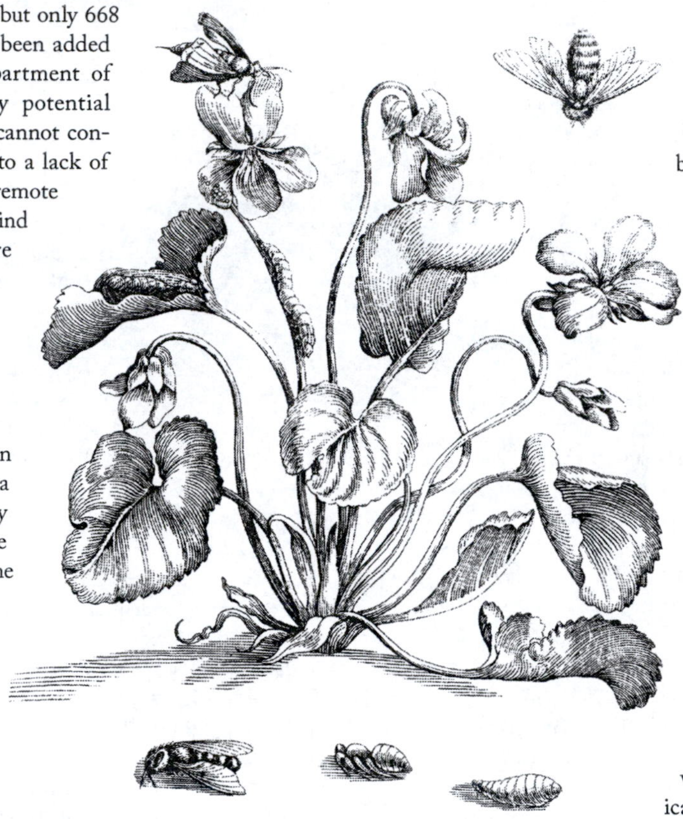

The current system in the United States for approving new transmission lines has contributed to the slow pace in overhauling the grid, although it is in desperate need of modernization. State and local regulators determine whether new transmission lines are needed, and focus on benefits to in-state ratepayers. Some states take into consideration renewables and other climate goals during transmission planning and siting decisions, but most states do not address these concerns. A project can fail if states argue that their ratepayers will pay for a project primarily benefiting customers in other states, as is the case with renewables projects connected to a grid that crosses state lines.

FERC traditionally had no jurisdiction over transmission siting, which was left entirely to the states. In 2005, a new Federal Power Act section expanded FERC's limited powers, allowing it to designate "national interest electric transmission corridors" and exercise "backstop" authority. This empowers FERC to preempt a state and issue a permit for the siting, construction, or modification of a proposed transmission line in a designated corridor if a state has withheld approval for more than a year. Federal courts, however, have rejected a number of FERC's attempts to overcome state resistance to transmission projects.

The American Clean Energy and Security Act climate bill, passed by the US House of Representatives in 2009, endorsed regional planning for new transmission lines and would have empowered FERC to review regional transmission plans. Among other goals, FERC would have been directed to approve transmission plans that "facilitate the deployment of renewable and other zero-carbon and low-carbon energy sources for generating electricity to reduce greenhouse gas emissions." With climate bills stalled, however, transmission siting remains largely an issue of state and local regulation.

Another significant barrier facing new transmission lines is deciding who will pay for them. Typically, ratepayers in the area of the project pay for new transmission lines through rate increases. This makes it uneconomical for utilities to invest in transmission projects that will benefit customers in other parts of the nation. A difficult issue is how the costs of future investments in the transmission grid will be shared differently, reflecting environmental and economic benefits that transcend the territorial boundaries of individual utilities.

Smart Grid Programs

Upgrading transmission lines is one component of the effort underway in the United States to develop a smart grid.

The current electricity transmission grid does a good job of what it was designed to do: one-way delivery of a product (electricity) that cannot be stored and must be consumed as soon as it is made. Despite occasional high-profile outages and blackouts, the system is generally strong and reliable. The term *smart grid* refers to a next-generation electricity transmission and distribution network that would provide many more capabilities through real-time two-way communication between consumers and utilities. A smart grid could drive much wider adoption of renewables and demand reduction, as well as improve reliability, operating efficiency, and grid resiliency to terrorist threats.

A smart grid would require many new ideas and technologies to achieve these disparate objectives. For example, it would rely on advanced meters that provide real time data between the customer and utility. When customers can tell how much electricity they are using and what it costs, they might be more inclined to buy "smart" devices like thermostats, clothes washers and dryers, microwaves, hot water heaters, and refrigerators that can interact with the smart grid. A Department of Energy study has found that these devices can reduce demand and energy costs (US Department of Energy 2009). Advanced meters could enable real-time pricing programs as well.

A major focus of the smart grid would be managing the two-way flow of electricity, enabling customers to generate or store electricity and sell it back to the grid during peak periods, when prices are highest. This would be a major incentive for deployment of small-scale solar, wind, and other renewable energy systems. Consumers could also store electric energy during the day in plug-in hybrid electric vehicles, and provide it back to the grid in off-hours. Managing the reverse flow of electricity will require sophisticated smart grid technology to overcome possible safety and reliability issues.

Developing a smart grid in the United States will entail massive improvements to the system currently in place and billions of dollars in funding. The federal stimulus package of 2009 allocated $3.4 billion for smart grid projects, but this is only a portion of the amount necessary. Development efforts are underway at both prominent American companies, such as IBM and Google, and grid technology startups, which have moved to create smart grid technologies (The Cleantech Group 2010). Simply establishing the rules by which a smart grid will operate is a daunting task, and a wide variety of technical standards for the smart grid are being developed in an effort led by a governing board at the US National Institute of Standards and Technology. In China, the State Grid Corporation's "Strengthened Smart Grid" plan calls for a similar effort oriented toward the construction of new transmission lines and the development of standards for a smart grid as well as the technologies needed for operation and control. A cooperative effort with the United States to foster development of these technologies is also underway.

Decoupling

An important regulatory mechanism to promote utilities' environmental programs is decoupling, breaking the link between sales and profits. Decoupling gives utilities incentives to adopt programs that encourage consumers to use less electricity. Rate cases continue as before, but price adjustments between cases allow utilities to recover revenue totals that are redefined as consumers use less electricity. As of 2009, seventeen states have decoupling mechanisms for individual utilities. Adjusted rates are still set as per-unit charges that give customers an incentive to reduce energy consumption, while enabling the utility to recover costs and revenue.

More frequent adoption of decoupling may facilitate progress toward reducing greenhouse gas emissions. Some call energy efficiency and DR programs the least expensive means for reducing electricity consumption and carbon emissions (United Nations Foundation 2007). Decoupling is itself not a means to address climate change. It does, however, allow a utility to recover its fixed costs even if consumption declines, and thus it may help implement the wide variety of carbon reduction policies being developed by states, regions, and the federal government.

Joel Barry EISEN
University of Richmond School of Law

See also in the *Berkshire Encyclopedia of Sustainability* Climate Change Disclosure—Legal Framework; Energy Conservation Incentives; Energy Subsidies; Environmental Law—China; Environmental Law—United States and Canada; Free Trade; Green Taxes; Investment Law, Energy

Further Readings

Averch, Harvey, & Johnson, Leland L. (1962). Behavior of the firm under regulatory constraint. *American Economic Review, 52*(5), 1052–1069.

Black, Bernard S., & Pierce, Richard J., Jr. (1993). The choice between markets and central planning in regulating the US electricity industry. *Columbia Law Review, 93*, 1339.

Brown, Matthew H., & Sedano, Richard P. (2003). A comprehensive view of US electric restructuring with policy options for the future (Electric Industry Restructuring Series, National Council on Electricity Policy). Retrieved September 15, 2010, from http://www.hks.harvard.edu/hepg/Papers/BrownSedano.pdf

Cavanagh, Ralph C. (1986). Least-cost planning imperatives for electric utilities and their regulators. *Harvard Environmental Law Review, 10*(2), 299–344.

Ceres. (2010). The 21st century electric utility: Positioning for a low-carbon future. Retrieved September 13, 2010, from http://www.ceres.org/Page.aspx?pid=1263

Chen, Cliff; Wiser, Ryan; & Bolinger, Mark. (2007). Weighing the costs and benefits of state renewables portfolio standards: A comparative analysis of state-level policy impact projections. Retrieved September 27, 2010, from http://eetd.lbl.gov/ea/ems/re-pubs.html

The Cleantech Group LLC. (2010). 2010 US smart grid vendor ecosystem: Report on the companies and market dynamics shaping the current US smart grid landscape. Retrieved September 27, 2010, from www.energy.gov/news/documents/Smart-Grid-Vendor.pdf

Commission of the European Communities. (2008). The support of electricity from renewable energy sources. Retrieved September 27, 2010, from http://ec.europa.eu/energy/climate_actions/doc/2008_res_working_document_en.pdf

Database of State Incentives for Renewables and Efficiency (DSIRE). (1998). State programs and regulatory policies report. Retrieved September 20, 2010, from http://www.dsireusa.org

Davies, Lincoln L. (2010). Power forward: The argument for a national RPS. *Connecticut Law Review, 42*(5), 1339.

Eisen, Joel B. (2010a). China's renewable energy law: A platform for green leadership? *William and Mary Environmental Law and Policy Review, 35*, 1.

Eisen, Joel B. (2010b). Can urban solar become a "disruptive" technology?: The case for solar utilities. *Notre Dame Journal of Law, Ethics & Public Policy, 24*, 53.

Estache, Antonio, & Martimort, David. (1999). Politics, transaction costs, and the design of regulatory institutions (World Bank Policy Research Working Paper No. 2073). Retrieved October 25, 2010, from http://papers.ssrn.com/sol3/papers.cfm?abstract_id=620512

Hendricks, Bracken. (2009, February). Wired for progress: Building a national clean-energy smart grid. *Center for American Progress.* Retrieved September 27, 2010, from http://www.americanprogress.org/issues/2009/02/wired_for_progress.html

Herzog, Antonia V.; Lipman, Timothy E.; Edwards, Jennifer L.; & Kammen, Daniel M. (2001). Renewable energy: a viable choice. *Environment, 43*(10), 8–20.

Lesh, Pamela G. (2009). Rate impacts and key design elements of gas and electric utility decoupling: A comprehensive review. Retrieved September 13, 2010, from www.raponline.org/Pubs/Lesh-CompReviewDecouplingInfoElecandGas-30June09.pdf

Pew Center on Global Climate Change. (2010). Homepage. Retrieved September 30, 2010, from http://www.pewclimate.org/

The Public Utility Regulatory Policies Act of 1978, Pub. L. No. 95–617, 92 Stat. 3117 (codified in scattered sections).

Rickerson, Wilson; Bennhold, Florian; & Bradbury, James. (2008). Feed-in tariffs and renewable energy in the USA—A policy update. Retrieved September 27, 2010, from www.wind-works.org/FeedLaws/USA/Feed-in_Tariffs_and_Renewable_Energy_in_the_USA_-_a_Policy_Update.pdf

Rossi, Jim. (2009). The political economy of energy and its implications for climate change legislation. *Tulane Law Review, 84*, 379–428.

Rossi, Jim, & Brown, Ashley C. (2010). Siting transmission lines in a changed milieu: Evolving notions of the "public interest" in balancing state and regional considerations. *University of Colorado Law Review, 81*, 705.

Shapiro, Sidney A., & Tomain, Joseph P. (2005). Rethinking reform of electricity markets. *Wake Forest Law Review, 40*, 497–543.

Shirley, Wayne. (2010). Mechanics & application of decoupling. Retrieved September 13, 2010, from http://www.raponline.org/docs/RAP_Shirley_PennsylvaniaDecoupling_2010_04_28.pdf

Stigler, George J., & Friedland, Claire. (1962). What can regulators regulate? The case of electricity. *Journal of Law and Economics, 5*, 1–16.

United Nations Foundation. (2007). *Realizing the potential of energy efficiency: Targets, policies, and measures for G8 countries.* Retrieved October 25, 2010, from http://www.globalproblems-globalsolutions-files.org/unf_website/PDF/realizing_potential_energy_efficiency.pdf

US Department of Energy (DOE). (2002). A primer on electric utilities, deregulation, and restructuring of US electricity markets. Retrieved October 25, 2010, from http://www1.eere.energy.gov/femp/pdfs/primer.pdf

US Department of Energy (DOE). (2006). Benefits of demand response in electricity markets and recommendations for achieving them. Retrieved September 13, 2010, from http://eetd.lbl.gov/ea/EMP/reports/congress-1252d.pdf

US Department of Energy (DOE). (2008). The smart grid: An introduction. Retrieved September 27, 2010, from http://www.oe.energy.gov/DocumentsandMedia/DOE_SG_Book_Single_Pages.pdf

US Department of Energy (DOE). (2009). Smart grid system report. Retrieved September 27, 2010, from http://www.oe.energy.gov/DocumentsandMedia/SGSRMain_090707_lowres.pdf

US Department of Energy (DOE). (2010). Demand-side management actual peak load reductions by program category. Retrieved October 25, 2010, from http://www.eia.gov/cneaf/electricity/epa/epat9p1.html

US Department of Energy (DOE) & Energy Information Administration (EIA). (2000). The restructuring of the electric power industry: A capsule of issues and events. Retrieved September 13, 2010, from http://tonto.eia.doe.gov/FTPROOT/other/x037.pdf

US Department of Energy (DOE) & Energy Information Administration (EIA). (2002). US electric utility demand side management (DSM) data 2000. Retrieved September 15, 2010, from http://www.eia.doe.gov/cneaf/electricity/page/eia861dsm.html

US Environmental Protection Agency (EPA). (2008). National action plan vision for 2025: A framework for change. Retrieved September 13, 2010, from http://www.epa.gov/cleanenergy/energy-programs/suca/resources.html

Index

A

B

C

Bold entries and page numbers denote article titles in this book.

D

E

Bold entries and page numbers denote article titles in this book.

O

P

R

S

T

Bold entries and page numbers denote article titles in this book.

U

V

W

Y

Image Credits

The illustrations used in this book come from many sources. There are photographs provided by Berkshire Publishing's staff and friends, by authors, and from archival sources. All known sources and copyright holders have been credited.

Bottom front cover photo is of fireflies (*Pyractomena borealis*) on an Iowa prairie, by Carl Kurtz.

Engraving illustrations of plants and insects by Maria Sibylla Merian (1647–1717).
Beetle, dragonfly, moth, and ladybug illustrations by Lydia Umney.

Front cover images, left-to-right:

1. *Waves, South Africa.* Photo by Rupert Jefferies.
2. *Vivid orange streamers of super-hot, electrically charged gas (plasma) arc from the surface of the Sun, 20 November 2006.* Photo courtesy of Hinode, JAXA/NASA.
3. *Bamboo.* Photo by Anna Myers.

Back cover images, left-to-right:

1. *Wind turbines in Somerset, Pennsylvania, USA.* Photo by Eric Vance, Chief Photographer, US Environmental Protection Agency, National Archives.
2. *General view of the city and the Atchison, Topeka, and Santa Fe Railroad, Amarillo, Texas; Santa Fe R.R. trip.* Photo by Jack Delano, Library of Congress
3. *Still image of Bangkok, Thailand, from video taken by the crew of Expedition 29 onboard the International Space Station.* Image courtesy of the Image Science & Analysis Laboratory, NASA Johnson Space Center. Still number ISS029-E-37074.

Pages VII, 161, and 170, *Water lilies.* Photo by Carl Kurtz.

Pages 1 and 70, *Azalea in the fall.* Photo by Anna Myers.

Page 7 and 12, *Connestee Falls, Brevard, North Carolina, USA.* Photo by Ellie Johnston.

Page 18, *Pink sedum,* Anna Myers.

Page 23, *Wildflowers, Saint-Gilles, France.* Photo by Ellie Johnston.

Pages 27 and 124, *General view of the city and the Atchison, Topeka, and Santa Fe Railroad, Amarillo, Texas; Santa Fe R.R. trip.* Photo by Jack Delano, Library of Congress.

Page 31, *Great blue herons.* Photo by Carl Kurtz.

Pages 35, 93, 124, and 154, *Still image of Bangkok, Thailand, from video taken by the crew of Expedition 29 onboard the International Space Station.* Image courtesy of the Image Science & Analysis Laboratory, NASA Johnson Space Center. Still number ISS029-E-37074.

Pages 38 and 89, *Rudbeckia 'Prairie Sun' flowers, Butchart Gardens, British Columbia, Canada.* Photo by Ellie Johnston.

Page 43, *Grasses.* Photo by Anna Myers.

Pages 48, 64, and 137, *Hot pool, Yellowstone National Park, Wyoming, USA.* Photo by Amy Siever.

Page 52, *Dam and reservoir, Spain.* Photo by Ellie Johnston.

Pages 60 and 76, *Vivid orange streamers of super-hot, electrically charged gas (plasma) arc from the surface of the Sun, 20 November 2006.* Photo courtesy of Hinode, JAXA/NASA.

Page 81, *Waves, South Africa.* Photo by Rupert Jefferies.

Page 84, *Wind turbines in Somerset, Pennsylvania, USA.* Photo by Eric Vance, Chief Photographer, US Environmental Protection Agency, National Archives.

Page 98, *Bamboo.* Photo by Anna Myers.

Page 104, *Prickly pear cacti and flowers, Red Rocks State Park, Sedona, Arizona, USA.* Photo by Amy Siever.

Pages 109 and 114, *Craggy mountainside, China.* Photo by Joan Lebold Cohen.

Pages 111 and 114, *Farm in Iowa, USA.* Photo by Berkshire Publishing.

Page 116, *Fossil shells embedded in rock at Castle Point, New Zealand.* Photo courtesy of morguefile.com.

Page 119, *Open mine.* Library of Congress.

Page 132, *Waterfall, Kent Falls State Park, Kent, Connecticut, USA.* Photo by Amy Fredsall.

Page 143, *Sugarcane land, vicinity of Rio Piedras, Puerto Rico.* Library of Congress.

Page 145, *Alpine field, Huaraz, Peru.* Photo by Ellie Johnston.

Page 147, *Poppy, Cotswold Hills, Gloucestershire, UK.* Photo by Amy Siever.

Author Credits

Introduction to Renewable Energy
by **Gernot Wagner**
Environmental Defense Fund

Algae
by **Juan Lopez-Bautista** and **Michael S. DePriest**
The University of Alabama

Aluminum
by **Greg Power**
Arriba Consulting Pty Ltd.

Cap-and-Trade Legislation
by **John C. Dernbach**
Widener University Law School

Carbon Footprint
by **Andrea Sarzynski**
University of Delaware

Coal
by **Zhao Ning, Song Quanbin,** and **Lian Ming**
Shanghai Bi Ke Clean Energy Technology Co., Ltd.
and **Xu Guangwen**
Chinese Academy of Sciences

Energy and Theology
by **Sarah E. Fredericks**
University of North Texas

Energy Conservation Incentives
by **Adrian DiCianno Newall**
Energy Consultant

Energy Efficiency Measurement
by **Horace Herring**
The Open University

Energy Industries—Bioenergy
by **Andrew Dane**
University of Wisconsin-Extension

Energy Industries—Geothermal
by **Alexander Richter**
Islandsbanki, Geothermal Energy Team; ThinkGeoEnergy

Energy Industries—Hydroelectric
by **André Lejeune**
University of Liège

Energy Industries—Hydrogen and Fuel Cells
by **Stefano Pogutz** and **Paolo Migliavacca**
Bocconi University
and **Angeloantonio Russo**
Parthenope University

Energy Industries—Natural Gas
by **Kevin F. Forbes**
The Catholic University of America
and **Adrian DiCianno Newall**
Energy Consultant

Energy Industries—Nuclear
by **Truman Storvick**
University of Missouri, Emeritus

Energy Industries—Solar
by **Michael Dale Harkness**
Cornell University

Energy Industries—Wave and Tidal
by **Stephen Turnock**
University of Southampton

Energy Industries—Wind
by **Kevin F. Forbes**
The Catholic University of America
and **Adrian DiCianno Newall**
Energy Consultant

Energy Labeling
by **Ornella Malandrino**
Salerno University

Energy Subsidies
by **Joshua P. Fershee**
University of North Dakota School of Law

Investment, CleanTech
by **Peter Adriaens**
Ross School of Business, University of Michigan

Investment Law, Energy
by **Peter Cameron and Abba Kolo**
University of Dundee

Iron Ore
by **Gavin M. Mudd**
Monash University

Lighting, Indoor
by **Heather Chappells**
Saint Mary's University

Lithium
by **Gavin M. Mudd**
Monash University
and **Steve H. Mohr**
University of Technology, Sydney

Materials Substitution
by **Asheen A. Phansey**
Dassault Systèmes SolidWorks Corp.; Babson College

Mining
by **Dirk Van Zyl**
University of British Columbia, Vancouver
and **David Gagne**
Berkshire Publishing Group

Petroleum
by **Charles A. S. Hall**
State University of New York, Syracuse

Polluter Pays Principle
by **Gerhard Roller**
University of Applied Sciences

Shale Gas Extraction
by **Robert W. Howarth**
Cornell University

Sugarcane
by **Shotaro Ando**
National Institute of Livestock and Grassland Science, Japan

Thorium
by **Casey Coombs**
University of Utah

Uranium
by **Casey Coombs**
University of Utah

Utilities Regulation
by **Joel Barry Eisen**
University of Richmond School of Law

This **BERKSHIRE** *Essentials* book was distilled from the

Berkshire Encyclopedia of Sustainability VOLUMES 1–10

Knowledge to Transform Our Common Future

In the 10-volume *Berkshire Encyclopedia of Sustainability*, experts around the world provide authoritative coverage of the growing body of knowledge about ways to restore the planet. Focused on solutions, this interdisciplinary print and online publication draws from the natural, physical, and social sciences—geophysics, engineering, and resource management, to name a few—and from philosophy and religion. The result is a unified, organized, and peer-reviewed resource on sustainability that connects academic research to real world challenges and provides a balanced, trustworthy perspective on global environmental challenges in the 21st century.

10 VOLUMES • 978-1-933782-01-0
Price: US$1500 • 6,084 pages • 8½ × 11"

"This is undoubtedly the most important and readable reference on sustainability of our time"

—Jim MacNeill, Secretary-General of the Brundtland Commission and chief architect and lead author of *Our Common Future* (1984–1987)

Ray C. Anderson
General Editor

Sara G. Beavis, Klaus Bosselmann, Robin Kundis Craig, Michael L. Dougherty, Daniel S. Fogel, Sarah E. Fredericks, Tirso Gonzales, Willis Jenkins, Louis Kotzé, Chris Laszlo, Jingjing Liu, Stephen Morse, John Copeland Nagle, Bruce Pardy, Sony Pellissery, J.B. Ruhl, Oswald J. Schmitz, Lei Shen, William K. Smith, Ian Spellerberg, Shirley Thompson, Daniel E. Vasey, Gernot Wagner, Peter J. Whitehouse
Editors

"The call we made in *Our Common Future*, back in 1987, is even more relevant today. Having a coherent resource like the E*ncyclopedia of Sustainability*, written by experts yet addressed to students and general readers, is a vital step, because it will support education, enable productive debate, and encourage informed public participation as we join, again and again, in the effort to transform our common future."

—Gro Harlem Brundtland, chair of the World Commission on Environment and Development and three-time prime minister of Norway